THE
Touchstone of
ARCHITECTURE

Copyright © 2017 Read Books Ltd.
This book is copyright and may not be
reproduced or copied in any way without
the express permission of the publisher in writing

British Library Cataloguing-in-Publication Data
A catalogue record for this book is available from the
British Library

Architecture

Architecture (from the Latin *architectura*, after the Greek *arkhitekton*, meaning chief builder) is both the process and the product of planning, designing, and constructing buildings and other physical structures. It is an incredibly important part of human existence – starting from the simplest aspects of survival, yet also functioning as a cultural symbol, a works of art, and as a means of identification of past civilisations.

Building first evolved out of the dynamics between needs (shelter, security, worship, etc.) and means (available building materials and attendant skills). As human cultures developed and knowledge began to be formalized through oral traditions and practices, building became a craft, and 'architecture' was the formalised version of this craft. In many ancient civilizations, such as those of Egypt and Mesopotamia, architecture and urbanism reflected the constant engagement with the divine and the supernatural. Many ancient cultures resorted to monumentality in architecture (think of the Pyramids at Giza, or the Parthenon at Athens) to represent symbolically the political power of the ruler, the ruling elite, or the state itself.

The architecture and urbanism of the Classical civilizations such as the Greeks and the Romans generally evolved from civic ideals rather than religious or empirical ones – and new building types emerged. Architectural 'style' developed in the form of the Classical orders. The earliest surviving written work on the subject of architecture is *De Architectura*, by the Roman architect Vitruvius in the early first-century CE. According to Vitruvius, a good building should satisfy the three principles of *firmitas, utilitas,* and *venustas,* translating as 'durability', 'utility' and 'beauty'.

Early Asian writings on architecture include the *Kao Gong Ji* of China from the seventh century BCE; the *Shilpa Shastras* of ancient India and the *Manjusri Vasthu Vidya Sastra* of Sri Lanka. The architecture of different parts of Asia developed along different lines from that of Europe; Buddhist, Hindu and Sikh architecture each having different characteristics. Islamic architecture began in the seventh century CE, incorporating architectural forms from the ancient Middle East and Byzantium, but also developing features to suit the religious and social needs of the society. In Europe during the Medieval period, guilds were formed by craftsmen to organize their trades and written contracts have survived, particularly in relation to ecclesiastical buildings. From about 900 CE onwards, the movements of both clerics and tradesmen carried architectural knowledge across Europe, resulting in the pan-European styles Romanesque and Gothic.

In Renaissance Europe, from about 1400 onwards, there was a revival of Classical learning accompanied by the development of Renaissance Humanism, which placed greater emphasis on the role of the individual in society. Buildings were ascribed to specific architects – Brunelleschi, Alberti, Michelangelo, Palladio – and the cult of the individual had begun. Leone Battista Alberti, who elaborates on the ideas of Vitruvius in his treatise, *De Re Aedificatoria,* saw beauty primarily as a matter of proportion, although ornament also played a part. For Alberti, the rules of proportion were those that governed the idealised human figure; 'the Golden mean'.

The notion of 'style' in the arts was not developed until the sixteenth century, with the writing of Vasari. By the eighteenth century, his *Lives of the Most Excellent Painters, Sculptors, and Architects* had been translated into Italian, French, Spanish and English. With the emerging knowledge in scientific fields and the rise of new materials and technology, architecture and engineering further began to separate, and the architect began to concentrate on aesthetics and the humanist aspects, often at the expense of technical aspects of building design. Around this time, there was also the rise of the 'gentleman architect' who usually concentrated on visual qualities derived from historical prototypes, typified by the many country houses of Great Britain that were created in the Neo Gothic or Scottish Baronial styles.

The nineteenth-century English art critic, John Ruskin, in his *Seven Lamps of Architecture* (published 1849), had a representative view of what constituted architecture. Architecture was the 'art which so disposes and adorns the edifices raised by men ... that the sight of them contributes to his mental health, power, and pleasure.' For Ruskin, the aesthetic was of overriding significance. His work goes on to state that a building is not truly a work of architecture unless it is in some way 'adorned'. Around the beginning of the twentieth century, a general dissatisfaction with the emphasis on revivalist architecture and elaborate decoration gave rise to many new lines of thought that served as precursors to Modern Architecture.

Notable among these schools is the *Deutscher Werkbund*, formed in 1907 to produce better quality machine made objects. Following this lead, the *Bauhaus school*, founded in Weimar in 1919, redefined the architectural bounds; viewing the creation of a building as the ultimate synthesis – the apex of art, craft, and technology. When Modern architecture was first practiced, it was an avant-garde movement with moral, philosophical, and aesthetic underpinnings. Immediately after World War I, pioneering modernist architects sought to develop a completely new style appropriate for a new post-war social and economic order, focused on meeting the needs of the middle and working classes.

On the difference between the ideals of architecture and mere construction, the renowned twentieth-century architect Le Corbusier wrote:

> You employ stone, wood, and concrete, and with these materials you build houses and palaces: that is construction. Ingenuity is at work. But suddenly you touch my heart, you do me good. I am happy and I say: This is beautiful. That is Architecture.

Architecture itself has an incredibly long and fascinating history. As long as humans have been around, we have needed places to live, and have sought ways to make these spaces beautiful and functional. As our societies continue to change, so does the architecture which reflects them. It is hoped that the current reader enjoys this book on the subject.

THE
Touchstone of
ARCHITECTURE

by

SIR REGINALD BLOMFIELD, R.A., M.A.

D.Litt., F.S.A., Hon. Fellow of Exeter College, Oxford

PREFACE

THE following Essays differ in subject, but they have this in common, that the problems with which I have attempted to deal have been approached from the point of view of architecture. Hence my title, 'The Touchstone of Architecture'. Since the middle of the last century the Arts have been the happy hunting-ground of the literary man. The artist is too busy with his own work, and though he alone knows ' the aims and ideals of Art ', if I may borrow Mr. Clausen's title, he is not trained to compete with the gladiators of the pen. The result is, that with the public that takes some little interest in art the gladiators have it all their own way. They amuse themselves with setting up a succession of altars to unknown gods in painting and sculpture, and with disquisitions on architecture which have little relevance to the conditions under which that art is, and has to be, practised. About one hundred years ago Stendhal said of his contemporaries, 'Heureuse la littérature si elle n'était pas à la mode', and one is sometimes tempted to think that the connoisseur and the critic are the most dangerous enemies of art, because they will not leave the arts alone to follow their natural and logical development, but insist, *ex cathedra*, on imposing formulas of their own invention. The suggestions which I offer in these pages are the results of my experience in the practice of architecture, and of observation of contemporary art—and art criticism. If my statements appear to be dogmatic, let me at once disclaim any idea of dogmatism. In the difficult region of art, where so much depends upon intuition and emotion, all one can do is to feel one's way, to aim, at any rate, at a sound working hypothesis. Since the break-up of tradition, the Arts have been like ships without a compass. They have set an adventurous course, for classicism, for romance, for realism, and symbolism, only to find that for want of proper bearings they have missed their port in the end. The voyage is longer than they thought, the demands and difficulties very much greater. We can profit by this experience, and if we have to curb our ambition we can at least make sure of our aim.

REGINALD BLOMFIELD.

New Court,
   Temple,
     March 1925.

CONTENTS

	PAGE
Preface	v
State-aided Training in Art in England	1
The Artist and the Community	22
Famous Men	37
The Outlook of Architecture (1913)	54
Atavism in Art	72
The Bridges of London (1815–1920)	94
The Tangled Skein. Art in England, 1800–1920	115
Greek Architecture	139
Christopher Wren	175
Architecture and Decoration	202
Off the Track. Some Thoughts on Art	223

STATE-AIDED TRAINING IN ART IN ENGLAND [1]

I HAVE been asked to address you at your annual meeting, and it has given me very great pleasure to accept the invitation, because though we differ as to the means, we are all of us working for a common end, the improvement of the art of this country, in the widest sense of the term. At the same time I approach the subject with some trepidation, because, though I venture to consider myself an artist trained in the art that I endeavour to practise and have had some little experience in teaching in our Academy Schools, I cannot for a moment claim the knowledge which you gentlemen possess of the methods of art training ; and if I appear to you to go wide of the mark in any of my criticisms and suggestions you will, I am sure, take it in good part as an honest attempt to clear the air. It is only by free and frank discussion of this very difficult problem of State-aided art training from every point of view, that we can hope to arrive at consecutive and well-considered policy in regard to it. I may say at once that, in my opinion, the policy is still to seek, and that, until we have such a policy thoroughly thought out in its main issues, all that we do is mere tinkering, the patching up of rents in a garment that never has fitted particularly well and is now splitting all round.

[1] An Address delivered to the National Society of Art Masters, in the Theatre, South Kensington, on 25th July 1912.

It is a matter of common knowledge that the position of artists in the modern State, at any rate in England, is not a satisfactory one. For very brief periods, such as the latter part of the seventeenth century and throughout the eighteenth century, artists have enjoyed some real recognition in this country, but the attention of the English people has been elsewhere, and artists as a body have never occupied the position of esteem and consideration accorded to their colleagues in France, at any rate since the days of Louis XIV. Modern conditions seem to be ever against them. As the civilized world gets older, the accumulation of hoarded works of art increases on the one hand, and wealth appears to concentrate in the hands of a small minority on the other. The members of this small minority appear to be animated by a frantic desire to outbid each other, and there has arisen a class of collectors who buy for rarity and little else. The old-fashioned connoisseur and collector bought to suit his own fancy, and more often than not on his own judgement. If he liked the work of a living artist, he bought it, often it is true at a moderate price, but still he helped to form a market for living art. But these colossal buyers, who pay in thousands for works for which the artist was paid in tens, are killing modern art by turning their back on it and leaving it to starve. There are other circumstances which accelerate the process. The free and immediate interchange of ideas and fashions tends to weaken national sentiment and pride of race and country, and we have to face the facts that in the modern community art is to a great extent a luxury, subject to constantly changing fashions, and that in the second place there is too much of it, and what there is is not good enough. The conclusion to be drawn, both from the facts of modern life and from the experience of history, is that the policy of *laissez-faire*, of leaving things to find themselves, is not enough in the arts, and that to give them a fair chance the State must lend a hand in their organization.

The ascendancy in the Arts which France enjoyed after the days of Colbert and Louis XIV was the result of careful nursing from the first. Colbert did not care particularly for the Arts, nor, with the exception of architecture, did Louis XIV, but both of them were sagacious enough to see that the admirable national ability of the French could be turned to the account of the State if it was systematically trained, organized, and protected. The whole of Colbert's educational policy was devoted to this protection of French art and industry. He deliberately set himself to their consolidation by artificial means, and though he was not always successful, such industries as those of the Gobelins, of Beauvais, and of Aubusson remain to this day as monuments of his work ; and the policy so initiated has been persistently followed in France, as when in the eighteenth century Marigny [1] established the school and manufactory of Sèvres. The first point, then, that I would make, without dwelling on it further,

[1] The Marquis de Marigny, Surintendant des Batiments, 1754–74, and brother of Madame de Pompadour; see my *Hist. of French Architecture, 1661–1774*, ii, p. 135, and vol. i, chaps. 1, 2, 3.

is that the market for the Arts under modern conditions in England is bad, and that there are two ways of meeting this, either the method of Colbert, frank and even ruthless protection, or the diminution of the supply in quantity and its improvement in quality.

Now how does the State in England deal with this problem of a failing market? So far it has not dealt with it; so far the problem hardly seems to have been realized, and the reasons are not very far to seek. It appears from the Historical Memorandum printed as an Appendix to the Report of the Royal College of Art Departmental Committees that we in England were just one hundred and seventy years behind the French in inquiring into the question at all. In 1835 a select committee of the House of Commons was appointed ' to inquire into the best means of extending a knowledge of the arts and of the principles of design among the people (especially the manufacturing population) of this country '. The first Government School of Design was established by the Board of Trade in 1837 ' with the avowed aim of encouraging the direct application of the arts to manufactures '. The instructors in this School fell out among themselves, and in 1852 the Department of Practical Art was established, which in the following year became the Department of Science and Art. The Department undertook to train Art masters, to furnish copies of models and to organize exhibitions and awards, the local committees undertaking on their part to provide schools and be responsible for finance. The next stage came with the establishment of payment by results, in

1862, and the establishment of 'Instruction in Drawing in Schools for the poor', as part of general education. In 1864 a Select Committee was appointed to inquire into the working of Schools of Art, and grants in aid of National Education in Art. A fresh classification was made, and this reorganized system, with modifications, has lasted to the present day. With its details I need not trouble you, as you are far more familiar with them than I am ; but I am informed by the Board of Education that there are now in England and Wales 223 'Schools of Art', 97 of which are open for eight or more morning or afternoon meetings a week, exclusive of Saturdays, and may be regarded as providing instruction for those students who are able to devote their whole time to the study of art, to the exclusion of any other occupation. There are 41,729 students, of whom 6 per cent. are full-time students, in the somewhat liberal definition of a full-time student as one who is engaged in a course of instruction contemplating his attendance for not less than twenty hours a week of instruction and of regular practice within the School outside the hours of instruction, and the average annual cost of maintenance of Schools of Art may be taken to be from £4 5s. to £4 10s. per student.[1] I gather that there would have to be added to this, at any rate, the cost of maintenance allowances awarded to students holding scholarships. In regard to control, the Board of Education do not attempt to dictate any rigid curriculum or framework of studies for Schools of Art, and the schools have been allowed to develop on the

[1] Written in 1912.

lines best suited to the needs of the locality. I may add that, up to a comparatively recent period, the principal officials at South Kensington, other than the Director of Art, were retired officers of the R.E. who, however eminent in science, were totally ignorant of art. The Inspecting Staff of the Board, consisting of a chief inspector and four subordinate inspectors, supervise the instruction, chiefly however with a view to seeing that the regulations in regard to the granting of State aid are duly observed; and it appears that the only means available of testing the work done in the Schools are (1) the 'personal examinations' conducted by the Board, and (2) the national competition for 'works of special excellence' executed by students in the recognized schools within the past twelve months. The Board also relies largely on 'adequate supervision on the part of managing committees of Schools of Art', and does all that it can to encourage local interest in the Schools. Stress has always been laid on the intimate connexion between Schools of Art and local industries and manufactures, and this may be regarded as the aspect of education in art which has received the greatest amount of official encouragement. Admission to the Schools appears to be open to any one who can pay the very moderate fees demanded, but the regulations on this point are obscure. In Part I, Clause 10 of Grant Regulations (1910), it is stated 'no student may be refused admission on other than reasonable grounds', but Clause 22A says the 'regulations' for admission must be such as to 'exclude from a course or class any student who from want of sufficient preliminary training or other cause is not qualified to take advantage of the instruction given in it'. As worded, the regulations might mean business, or might merely be intended to exclude loobies. I take it, in any case, that the standard of admission is a low one.

You must excuse this recapitulation of facts with which you are all familiar, but it is necessary to my argument. We have now to look at the results of this system. They are given with alarming precision in the Report of the Departmental Committee on the Royal College of Art issued in 1911, par. 27, 'on the training of designers at the College'. The Committee, referring to the expectation that designers trained at the College, and under our State-aided system, would be eagerly snapped up by manufacturers, reports ' all the evidence which we have been able to collect from many and divers sources of information has brought us to the regrettable conclusion that such is not in fact the case'. The regrettable conclusion, which unfortunately there is no reason to doubt, proves that the well-meant efforts of the last seventy years have resulted in failure. I mentioned above that the general attitude of the Board points to the intimate alliance of Art and Manufactures as the principal object of our State-aided training, yet here we have the Departmental Committee stating in terms that this is precisely what it has failed to bring about ; and it has failed because from 1835 onwards the authorities have never been clear in their own minds whether their object was educational or technical ; whether, that is, they wished to improve the aesthetic intelligence of the English people, and as they put it, 'to extend a knowledge of the arts and of the principles of design among the people'—or whether their object was to 'encourage the direct application of the arts to manufactures'. As a matter of fact, they have wished to do both, but they have allowed considerations which properly belong to one aim to affect the other. Colbert, most clear-headed of reformers and organizers, was under no such illusion. His avowed policy was to increase the wealth of France by the development of her manufactures. Where these manufactures involved the arts, he provided for the arts in a liberal and most intelligent manner, for he provided Schools of Art in direct communication with the subsidized industries such as the 'École de la manufacture nationale des Gobelins', or the 'École des Arts décoratifs de la manufacture de Beauvais'. The French School at Rome was an extension of the same central idea, first that, in regard to the State, artists were to be trained, and thoroughly trained, in order to add to the wealth and prestige of the State—all the work, for example, done at the School of Rome was regarded as the property of the King—and secondly that, in regard to the artist, all who intended to practise the Arts and Crafts should receive a grounding in the fundamental methods of drawing and modelling as well as in the technical processes, and that training in both should proceed, as far as possible, hand in hand. The designer must learn to draw, and he must be brought into intimate familiarity with the workshop, if he is to be of practical use to the manufacturer, but both elements of his training were considered essential, and undue stress was not to be laid on one at the expense of the other.

Scarcely less disquieting than this avowed failure to bring the arts of design into touch with manufacturers are the statistics of the National Competition. The Regulations refer to the National Competition as organized ' for works of special excellence '. That is, the works sent up by the schools are supposed to be selected works in the first instance. Yet in 1910, out of a total number of 13,097 works submitted in this competition, I am informed that 9,524 were either marked with a blue cross as too bad to be sent up, or were rejected at the first scrutiny as utterly out of the competition, that is, some 72 per cent. of the total were really bad. I gladly admit, as an examiner of some experience, that excellent and most promising work is also submitted from time to time, but this only adds force to the contention which I shall put before you, that drastic reorganization of our methods of art training is wanted. What becomes of the 72 per cent. of these young men and young women who have spent their own time and that of their instructors, and cost the tax-payers a considerable sum of money in the attempt to learn accomplishments for which they are evidently not qualified? It is to be hoped that they then and there abandon the attempt, but meanwhile they have wasted time and money which would have been better spent in learning something that they could do. More than that, these weak students are positively harmful to Schools of Art. They lower the standard and drag down the level, and they waste the masters' time and skill which would be better spent on really qualified students. It is a disquieting thought that after all this effort there is nothing for them to do, and that all this effort might have been saved had the aim of our State Art training been more clearly defined. The actual position appears to be that there exists on the one hand a large and costly apparatus of training, and on the other hand a poor and diminishing market for the wares that those so trained have to offer. The rich collector buys elsewhere, the manufacturer turns the cold shoulder on the designer, either because he is unpractical or because being trained too much on theory he is slow to handle the taste of the time, and though he may be right in doing so, that is not the idea of the manufacturer who is there to sell. Meanwhile, our Art Schools are steadily turning out large numbers of art students for whom there is nothing to do.

In making these observations I must not be supposed to be criticizing either the Board of Education, who administer with great ability a faulty system handed on to them by their predecessors, or the Art masters, a devoted body of men who have to carry on their work under conditions which for one reason or another must, I believe, lead to grievous disappointment. What I am hoping to do is, to clear the air by bringing the facts of the problem together, in order that by the full consideration of these facts we may arrive at some general policy of State-aided training in the Arts, under which each branch of training will fall into its appointed place. If the system is wrong to start with, no amount of patching will improve it.

The first question one would ask is, what is the end and object of this training? There appear to be several answers in the field. 'Fine Art'; the development of applied Art as opposed to Fine Art; the technical training of the artist and the craftsman; the education of the public in general and of the amateur in particular; and lastly, the preparation of the Art master. Mr. Burridge, in the excellent paper that he read to the Art masters last year, said that ' the functions of the School of Art are of a very diverse and exacting nature '. I would add that, as enumerated above, to a large extent they neutralize each other. If the training is to include the amateur, it is not adapted to the professional. If it is to be devoted to the education of the public, it must deal with matters of general culture which are outside the scope of the practising specialist. The Board of Education appear to incline, though not very decisively, to the view that the main object of this training is craftsmanship. Mr. Burridge put it very well : ' The aim of the School of Art is to make the artist a better craftsman and the craftsman a better artist.' Elsewhere he says : ' Craftsmen, designers for manufacturers, and workers at manufactures should be encouraged to train at a high artistic standard and, as far as opportunity permits, be given a sound artistic training as the basis of their specialized training.' These two statements appear to me to sum up correctly what should be the aim and policy of the training to be given in State-aided Schools of Art.

If this position is accepted without reservation, certain conclusions will follow from it which will involve the reorganization of the present system.

Let us consider the student first. The student will be trained to become a working artist or craftsman, that is, one whose qualifications natural or acquired enable and entitle him to take up one or other of the Arts or Crafts as the calling of his life. This enables us to clear away two parasitic developments of the Art School, the idea that one of its objects is general culture and education, and that it is part of its business to provide for the amateur. To both of these views, and they hang together pretty closely, I am strongly opposed. That some sort of rudimentary training of eye and hand should form a part of general education is admitted. What I contend is that it should be kept within its proper limits, dealt with, as in Scotland, in primary schools, and not overlap into the specialized instruction of the Art School. It is most dangerous to give any encouragement to the idea that because a boy or girl has dabbled with drawing or painting in their early youth they are therefore destined to become artists, and the training given in this early stage should be, as in the main I believe it is, handled from the point of view of mental training and the cultivation of powers of observation and analysis. I shall come later on to the methods of admission to the Art School. The idea that the Art School exists for the amateur is, I hope, now exploded. Anyhow, I contribute my parting kick. One has heard of Art Schools in towns where a special industry, such, for instance, as a carpet manufactory, exists, and where there was a young man with long hair and a velvet coat to teach the young ladies of the place drawing and painting, but no sort of effort was made to bring the School into relation to the craftsmen working in the manufacture on which the town relied for its very existence. Such a case has only to be stated to show the abuse of the system ; but it ought to be a definite condition of every Art School that it exists for real business, that is, for efficient and effective work. The Arts are a very charming amusement for amateurs, but the training that the amateur desires should be provided by private enterprise.

Another conclusion, again limiting the field, will follow. We should have it definitely laid down for us whether or not these schools are to deal with ' fine art ' as it is rather crudely called, by which I understand the three central arts of painting, sculpture, and architecture. The tendency of the Board, so far as I understand it, is to concentrate the work of the schools on industrial art, that is, on what for this purpose we may call craftsmanship, with the important proviso that the craftsmanship should be founded on a training in drawing and modelling as the basis of pretty nearly all the arts. If this position is accepted, the specialized study of painting and sculpture, apart from this preliminary training in drawing and modelling, would be left over to the private enterprise of artists. Architecture must be so left over in any case, owing to the difficulty of obtaining competent practical teachers, except in a very few schools. Those callings such as dentistry, cycle-making, and the like which have somehow managed to get themselves included in the curricula of schools of art would, I take it, resume their place in the technical school.

The schools, then, will be for the working artist craftsman. How is he to be admitted ? The conditions laid down by the Board of Education are to some extent contradictory, but in any case appear to be very loose, and to make admission mainly a question of fees. I suggest that no student ought to be admitted to the school unless he submits work proving that he possesses exceptional ability for the Arts, and in addition to these testimonies of study there should be some test of work done in examination in the school, showing that the competitor actually has his skill at the end of his fingers. We do not want, nowadays, in the Arts, the laborious practitioner ; there is room for the man with real gifts, but not for the other.

Now comes the all-important question of the schools and the training that can and should be given in them. I understand that there are in England some 223 Schools of Art. The subjects taught appear to range from drawing to dentistry. It is a little difficult to see the relevance of confectionery, bicycle manufacture, dentistry, gardening, and mining to a School of Art ; but they are all returned to the Board by one school or another as amongst the crafts in connexion with which special courses of art instruction are given. At one school, for example, dentistry is taught together with painting and architectural drawing. The percentage of full-time students in this school is 0·6, though it claims to rank as a ' full-time ' school. A second full-time school deals with building, engineering, and cabinet-making. Its percentage of full-time students is 0·5 and it possesses one solitary student struggling with advanced drawing.

Another school does better with 1·6 and deals with architecture, confectionery, sign-writing, and design. This, too, is a School of Art and is not part of a technical school. In addition to its one full-time student in advanced drawing, it also possesses a student in architecture, possibly to keep the confectioner in countenance.

These facts in regard to the Art Schools of this country are rather surprising, and they all point to the depressing conviction that, so far as the Arts are concerned, many of these schools are attempting a task for which they are in no way qualified or equipped, if we are to accept Mr. Burridge's proposition that the aim of these schools should be to make the artist a better craftsman, and the craftsman a better artist. How can one Art master possibly teach architecture, confectionery, and sign-writing ? Where does the Art master come in in the teaching of ' gardening, plumbing, and building ', or again in ' engineering, building, and mining ', or ' in carriage building, cabinet making, and building ', or ' engineering, building, and cycle manufacture '. To say the least of it, there appears to be some cross-classification here. The kind of drawing required for such industries should not be classified as art instruction at all. But in those many other schools where some art or another is taught (and I am not now referring to the large, well-equipped schools, but to the smaller schools where the master is single-handed) how can one man, however zealous, teach efficiently more than one art ? We artists who have spent the best years of our lives in trying to learn our art know perfectly well that the only way to master an art is to concentrate on that one art ; but here are men on short training set to teach painting, modelling, architecture, and two or three odd crafts as well. The result is inevitable; it is the blind leading the blind.

I suggest that the only effective way of dealing with the difficulty is, on the one hand, to limit the functions of the smaller and less-equipped schools, and, on the other, to develop the resources and staff of selected central schools. Mr. Burridge, whom I quote again, because he has put the matter decisively, insisted that students should be ' given a sound Art education as the basis of their specialized training ', and that principle applied with the necessary conditions and limitations is, I believe, the solution of the problem. The training provided in the smaller schools should be limited to drawing and modelling, and all craftsmen should be put through this course in some shape or other. To carry his training further he should be drafted off to one of the central schools, entrance to which should again be subject to some test and proof of the student's capacity. The teaching of the crafts and of designing for the crafts, which can only be taught to any purpose in connexion with the workshop, should be limited to the main central schools, and in the case of special crafts, to schools where there exists in the immediate neighbourhood an established industry in those crafts.

Lastly, I come to the teacher. Here I am skating on thin ice, and I shall only state my conviction that the best teacher in any art or craft is the working artist or craftsman, the man engaged in the actual exercise of the art that he has made his own, and who comes back to give the students the results of his hard-won experience. The teacher should not only be permitted, but should be encouraged, to continue the active practice of his art, both for his own sake and for the sake of his students. If this position is accepted, the test of capacity we shall look for in any who propose themselves as teachers will not be those dreadful mechanical drawings of architectural detail and the like that appear year after year at South Kensington, but the record of the work that the candidate has done ' on his own '. The result of the former regulations of the Board in regard to candidates for Art mastership is, that boys and girls set out from the first to qualify themselves for this profession, and they qualified themselves by a narrow, inadequate, and stereotyped course of study, which must effectually suffocate any latent instinct for the Arts. These methods recall the gerund-grinding of a past generation of teachers who used to make one regard the *Iliad* and the *Aeneid*, not as magnificent poems, but as textbooks of grammar. The methods by which the Art teachers' certificates were arrived at were even worse, because they did not even mean that grammar was mastered. If all this paraphernalia of Art teachers' qualifications were cut away, the masters of Art Schools would devote themselves to the training of their students as artists and craftsmen, and the students could enter with a whole heart on this training, freed from paralysing regulations which cramp their artistic impulse at every step. I do not want to be misunderstood on this point; what I am proposing will not make it easier to become an Art master or teacher, but much more difficult, because it will be necessary to prove real capacity as an artist instead of obtaining a merely mechanical pass. My object is to raise the status of the Art teacher by aiming at a higher standard of efficiency and by providing him with really qualified students in place of amateurs and dullards. It ought to be clearly understood that for positions solely affecting the Arts and Crafts, only artists and craftsmen are eligible. It is the exceptional, not the average, man that we want, and we shall not round the *impasse* that faces us by spreading ourselves out on the floor and trying to make things easy for everybody.

I must apologize for the length of this paper, but the subject is difficult and delicate and must be considered as a whole and in regard to social conditions. My points, briefly, are these. The artist and craftsman are in a precarious position, and their work is not in demand owing to the accumulation of works of art, and the particular direction given to connoisseurship by wealthy modern collectors. The State has failed to deal with this problem, it has on the contrary aggravated it by letting loose on the market large numbers of imperfectly trained artists, and this imperfect training is due to the absence of clear principle or policy in regard to the object and limits of State-aided training in art. The result is a large supply for which there is no demand. To meet these conditions we should definitely accept the position that the object in view is to produce really competent artists and craftsmen, men who will rank among the productive assets of the country.

This will clear away the confusion between general artistic education and specialized instruction and limit our training to the latter. It will, for this purpose, wipe out the amateur, and candidates for admission to the schools should have to give proof of exceptional capacity. The schools should be reduced in number, the smaller schools limited in function and be made preliminary and subordinate to a few central schools, which should be developed both as regards equipment and personnel. The instructors in these schools should be selected for proved ability in their art and not for their successful negotiation of certain mechanical tests, and as is done already in some of our more advanced schools, these men should be allowed the opportunity of practising their arts, giving for example half their time to the school and half to their private work, because only in this way can the teacher keep his art vital, and continue in touch with young and enthusiastic students.

I need hardly add that the limits of my paper do not allow me to go further into detail, but I would suggest, as a type of excellent organization, the system in use in Scotland. Here, instead of 218 or 223 schools unrelated to each other, they have three central schools for the areas into which Scotland is divided: Edinburgh, Glasgow, and Aberdeen. All other schools come under the general educational system, that is, the Art training given is only an extension of the teaching of drawing given in the primary school. These subordinate schools are under the control and direction of the main school (Glasgow, or whichever it may be) of the area in question, and the training given in them is limited to drawing and modelling, and is not carried further than the antique, no life class being permitted. Should the student wish to pursue his training further, bursaries and prizes exist for sending promising students to the central school. In its main features this seems to me an excellent system. It combines a real organization and correlation of all the Art Schools with the minimum of interference from official head-quarters. Each area, within limits, makes its own arrangements ; there is no official syllabus, this being left to the local centre, but on the other hand the local centre exercises a real control within its own area. There seems to me one weak point in the system : the keystone to the arch seems to be wanting. The primary school leads on to the continuation school, but there the system stops, and we are left with three central schools, of equal status, but necessarily varying standards and ideals. What is wanted is one great School of Art as the final objective of all the schools of Great Britain, a school to which entrance would only be possible by competition of the severest kind, because it would be a competition only open to the best of the students in the central schools, and these schools themselves would already consist of picked men. And as the prize of this school we should have our Prix de Rome.

As you are aware, the School of Rome is now an established fact, and we may hope to have there some day an Institution not wholly unequal to the famous school of the Villa Medici, and with the establishment of this final school our educational system will be complete. Its establishment in the near future is now a pressing necessity. Quite recently, through the courtesy of the authorities, I was able to inspect the ' École des Beaux Arts ' in Paris. I need say nothing as to that great institution; its prestige is world-wide, and yet, for several reasons, it is not wholly suited to the English student. We want some such school as this in England if we are ever to revive the traditions of our own great masters of the past, if we are to do justice to the latent capacities of our race. But in order to lay its foundation surely, we must begin at the bottom, we must overhaul our present methods, we must reorganize the system from top to bottom, and we must take as our motto, Quality not quantity.

June 1912.

THE ARTIST AND THE COMMUNITY[1]

I HAVE been invited to address you to-night on the occasion of the annual prize-giving. It is an honour and it is a great pleasure to me to have this opportunity of addressing you and of visiting your school, placed as it is in the centre of a very important industry, where the training given in the school, and the staple manufacture of the town, are able to act and react on each other in a manner which I believe to be vital to the teaching of the industrial arts. The problem of making our training in those arts really efficient is much in the air just now, and certain recent remarks which I have ventured to make on this subject have, I fear, been misunderstood by some of my critics. I am not less anxious than they are to promote the teaching of the Arts and Crafts ; I realize, as they do, their importance in the life of the community ; but as a practical designer myself I am convinced of the vital necessity of bringing those Arts and Crafts into touch with facts, of insuring that the design taught in our schools is not treated solely as a matter of draughtsmanship and paper work, but recognizes to the full the conditions and limitations of the particular art and craft with which it deals. In the industrial arts, as in architecture, draughtsmanship is not an end

[1] An Address delivered on the occasion of the distribution of prizes to the students of the Technical School of Art, Sheffield, on 21st February 1913.

in itself, but a means to an end, namely, the realization of a definite result in the materials of the art. I am convinced that, to ensure that there is no waste of labour, no futile effort, design and process must go hand in hand. The designer must have ever present in his consciousness the necessary limitations of his design in practical conditions ; if it is for textiles, in processes of weaving and dyeing ; if it is for wall-papers, in the possibilities of printing ; if it is window-glass, in the properties of glass, and its behaviour under firing ; if it is metal-work, in the multifarious methods of manipulating metal. It is the duty of the designer to lend himself to the practical methods of his craft, not to fight against them, but to draw his inspiration from their necessary conditions. As for the other side of his training, the patient labour of drawing or modelling, the study of great examples, the steady research into Art in general, and his own art in particular, its object is on the one hand to furnish the designer's mind with material on which it can work, a rich store of ideas and ever-present consciousness of the beautiful things that have been done by craftsmen in the past ; and, on the other hand, to provide him with the means of expression for his ideas, so that the flow of his invention is not checked by incompetence of hand.

All thoughtful observers have long recognized the necessity of this intimate alliance between the designer and the workshop. They should never be out of touch with each other, and design in industrial art has only value in relation to the terms of its realization. In other words, the test of the value of a design is the result translated into actual materials. That was the motive that prompted William Morris in forming his famous company. His example was followed by others of a younger generation, and inspired all the best work of the Arts and Crafts Society in the brilliant days of its youth. It is an axiom of training that has, I fear, been somewhat generally overlooked. South Kensington, a generation ago, carried with it ominous suggestions of design on abstract principles, of design based on formulae of conventionalized art, which has been to a large extent responsible for the gulf fixed between the manufacturer and the designer.

It is refreshing to come to a great centre, such as Sheffield, the home in the eighteenth century of one of the most beautiful of our English crafts, where design and craftsmanship can check and reinforce each other in a manner denied to those less-favoured places where there may be schools of design without the local industry, or the local industry without the school of design. Apart from the expense, I do not think it possible for the State to set up school workshops that can have the same vitality, the same intensity of purpose as those workshops and manufactories that are run on a business basis. And in the historical examples, such as the factories of Gobelins, or Beauvais, where workshop and school of design were established together by the State, the aim of Colbert, who established them, was the double aim of enriching the State and of improving the Art of the country ; the one not less than the other. I do not mean by this that the workshops attached to our Art Schools are to be closed down. What I do contend for is that the instructors in those workshops should be working artists and craftsmen, men who know, at first hand, what they are teaching. You have, then, in Sheffield, the right condition, as it seems to me, for the double purpose of a really efficient training of your artists and craftsmen, and for maintaining or raising the artistic standard of your principal industry. Your artists and craftsmen will gain by the discipline of facts in your workshops, and I venture to express the hope that your manufacturers will avail themselves of the trained and tempered skill that the artists have so acquired. For there are two sides to industrial art. I have laid great stress on practical knowledge as a vital element, but that does not exhaust the question. There is still design, the idea for the expression of which this practical knowledge is necessary. It is in your Art Schools, where Art is studied in a wider aspect, and by your Art masters and instructors, that this might be taught, and I venture to suggest that those great industrial organizers who have no leisure for this thorough study of Art should look for their designers in the schools of their own country. A prominent place is rightly assigned to silversmiths and metal-workers in the syllabus of your school, and the technical instruction in each class is given by practical craftsmen. I do not imagine that you gentlemen who control this great industry in Sheffield are likely to overlook such very exceptional resources. I have to-day had the pleasure of going round your Schools with Mr. Jahn, and have noted in the silversmiths' school some works in copper, simple in form, excellent in design and workmanship, which seem to me more worthy of encouragement than the highly ornamented work which has succeeded the great age of Sheffield plating.

There are, however, certain economic aspects of the present condition of the Arts in this country to which I would invite the careful attention of those of you who contemplate the career of the artist or the craftsman. If the picture which I present to you is not quite what you could wish, let me recall to you the fate of Ahab at Ramoth Gilead, the story of the false prophets who prophesied to Ahab, ' Go up to Ramoth Gilead and prosper ', and of Micaiah the son of Imlah, who told him the bitter truth, ' I saw all Israel scattered upon the hills as sheep that have no shepherd '. It is to enable you to avoid the fate of those who were scattered upon the hills like sheep that I feel it my duty to put before you what I believe to be the true state of the case, however unwelcome it may be.

The formidable fact that you have to reckon with is the present economic condition of the Arts in this country. It is admitted on every hand that it is far from satisfactory ; to put it bluntly, there is not enough work to go round. It is not merely that the supply exceeds the demand, but that the demand is for something different from what the average Art student is trained to supply ; and though I do not doubt that there are artists in this country who are capable of giving the very best in every art and every craft, yet they are inevitably few and far between, too often lost in the crowd, and deprived of the opportunity of giving of their best. Half the sensationalism and false appeal of modern art is due to this struggle for opportunity, to the temptation, which only strong artists can resist, to do anything to get out of the ruck. The condition of things is altogether different to what it was one hundred or one hundred and fifty years ago. From the point of view of the artist wanting his chance, society was better organized, because it was more compact and moved in certain well-defined grooves. The habits of the community, its demands and its resources, were well known. If a man wanted his portrait painted, he knew where to go : to the local painter in the provinces, if he were a man of modest means ; to Mr. Romney, if he wanted to be in the fashion ; to Mr. Gainsborough or Sir Joshua Reynolds, if he did not mind their price. There was no photographer to poach on the preserves of the painter, no dealers (or if there were any, they were dealers on a much more modest scale) to fill the rich man's gallery with old masters. People lived their lives in a restricted circle, quietly and without excitement, with leisure to concentrate on those aspects of life that pleased them, with a traditional standard of taste and refinement, little influenced by passing fashion, and with a critical sense of good and bad in art, which I believe to have been very much keener and sounder in the eighteenth century than it is to-day, in spite of the privilege we enjoy of having the whole theory of art expounded to us by the young lions of the Press.

Nor was this attractive state of things confined to London. In an able review, in the *Morning Post*, of Mr. Bradbury's fine history of old Sheffield Plate, the writer called attention to Sheffield Plate as ' one of the branches of art most perfectly typical of the singularly delicate and sane taste of the eighteenth century '. It was typical also of the active artistic life of the provinces ; and those of us who have studied what are called the minor arts of England in the eighteenth century have noted again and again what admirable work was done by the local cabinet-maker, the potter, the smith, the carpenter, and the joiner, in places far away from the influence of London, undeniable evidence of a real and complete vernacular art, that is an art in which all classes of the community took an interest, the man of small means as well as the man of great, an art that was understood by the people and loved by them. Yet in those days there were scarcely any exhibitions, not many dealers, and in the modern sense no Art Schools.

The fact is that we have to reckon with completely altered conditions. We have substituted quantity for quality, and though there is much individual artistic ability, there is no vernacular art. If, for instance, in the eighteenth century a man had ordered a new front door from his local builder, he would have known quite well what he would get, columns and pediment of correct proportions, raised and moulded panels, and some delicate glazing in the fan-light. Nowadays, if he did the same thing, he would have no sort of idea what to expect. Each order in the least off the beaten track is an experiment, and therefore people are slower to give them. There is no steady and constant demand, no set of thought in the same direction, by means of which the fulfilment of any well-defined want can attain its finest expression. The result of this is that bewildering succession of fashions borrowed from all times and all countries, which are the despair of the designer who has something to say of his own. Then there are the dealers who set the fashion in Old Masters, those princely tradesmen who have become the great financiers of art, and who have introduced to our modest civilization the prodigious prices and the reckless competition of Wall Street. Picture-dealing has become a lucrative business. Not unnaturally, fortunate dealers make the most of it, and to a very large extent control the market. If, for example, they have overstocked a modern artist, it is a simple matter to organize a boom in that artist. Mere studies are produced as finished masterpieces, and another red herring is dragged across the scent. The fashion for beautiful things is an important element in any complete culture, but when that taste is perverted into enthusiasm for rarity, with its corollary of costliness, it becomes quite a different thing, with little relation to the Arts, indeed injurious to their right appreciation, because it diverts attention from the true function of art, the selection and interpretation of beauty in its widest sense. Indeed, from this point of view, one of the worst traitors in the camp of art is the modern type of connoisseur who is mainly occupied with ' provenance ' and appraises works of art by collectors' values, without regard to their purpose as the expression of emotion and individuality. Age, of course, gives an added value to beautiful things, because, besides their intrinsic beauty, it invests them with a wealth of associations, derived from the fact that they express the ideals of past generations, that they have survived the vicissitudes of time, and that in the lapse of centuries they have acquired a certain mellow dignity denied to younger rivals. But when this feeling is exaggerated into a demand, not only for the literal reproduction of bygone art, but even for the simulation of the effects of age, it becomes the merest sentimentality, one of the most insidious enemies of the art of our own generation.

Lastly, there is the motor, and all that it means. The pace of life, the excitement, and a scale of expenditure that leaves little over for the encouragement of living artists. What a writer in *The Times* described as 'the motor standard' has materially changed the outlook on life in the last twenty years. Since the beginning of the nineteenth century our public has never shown a superabundant inclination for the Arts. Now it appears that what little attention it was ready to pay them has either been diverted to other tastes, or has settled into a groove of old masters, and archaeological revivalism. And on the top of this has come the widespread change in English social life, brought about by a change of locomotion which, for the purposes of ordinary life, has radically altered our ideas of space and time. In making these observations, I am making no criticism on motors, or dealers, or connoisseurs, but merely calling attention to certain facts of modern life, as I believe them to be, which make the economic situation to-day rather desperate for the artist who hopes to live by his art. The changes I have noted are a serious handicap to artists, because they are driven to violent self-assertion in order to get out of the crowd. Exhibitions are a severe trial to sensitive artists. They are compelled to scream, in colour, in sculpture, in building, because if they do not they are shouted down by less scrupulous competitors, and yet without these exhibitions how is an artist to get himself known? It is not very easy to see daylight for the Arts in all these difficult situations.

I fear there is no escaping the conclusion that the external conditions of modern life are at this moment unfavourable to the practice of the Arts. How about the artist himself? What, from the nature of his calling, is his position in the community? Here, again, I am unable to prophesy smooth things. I am not referring to that general training of the artistic sense which belongs to general education and, properly handled, is an important part of it, but to the artist and the craftsman who make Art the work of their lives—and I think here there has been a misunderstanding which has led to many a grievous disappointment. As the result of much meritorious effort on the part of educationists and philanthropists, we have come to look upon Art as one of the ordinary industries of life, one that can be taught in the same terms, and by the same methods, as matters of general education, on the one hand, or of technical instruction on the other; and that can be mastered by anybody, given a reasonable intelligence and steady application. It must be a comfortable view for the statesman and the philanthropist, because if it were so, all that would be necessary would be to provide a sufficient quantity of well-equipped and well-organized schools, and the result would be, if not a nation of artists, at least satisfactory performance on the part of all who thought fit to take up the calling of the artist. It is a view that has been acted on, with touching confidence, in this country, with the result that we have, at this moment, numbers of artists and craftsmen for whom there is little or nothing to do. I believe myself that there is a dangerous fallacy at the root of this view. It is that the Arts can be taught and learnt like anything else. Now I am not going to rely on the good old tag : 'Poeta nascitur, non fit ', but I put it to you that the Arts cannot be taught like anything else, because the subject-matter and the qualities involved are not normal but exceptional. I am not suggesting, like Dr. Nordau, that all artists are lunatics ; what I do say is that they are only artists in so far as they possess in an uncommon degree powers of observation, invention, imagination, of visualization and expression. You can teach a boy the grammar and elements of Latin or Greek, modern languages, mathematics or the rudiments of science, because all these are matters of exact knowledge, and the medium of expression is the written and spoken word, open to all who can read or write. But the medium of expression in the Arts, both for the teacher and the student, is not, or is not only, the written or spoken word. The artist expresses himself in his drawing, painting, sculpture or architecture, or in any of the crafts, by his actual handiwork ; he does not describe what he is about, he does it. It is there that he expresses himself freely and fully, not in the written or spoken word. And for the teacher in the Arts, the most efficient and valuable teaching that he gives is given by actually showing his students where they are wrong in method, how the thing ought to be done in actual execution. All of us who have studied under good teachers know that a mere hint in words, demonstrated by the teacher with brush, or pencil, or on the clay itself, is what is wanted by the artist and the craftsman for the special purpose of his training as an artist. My point is that in the arts and handicrafts the appeal is made through the eye to the aesthetic sense, or whatever we are to call it, and that therefore the manipulation of the actual material which forms the medium of expression becomes of supreme importance both for teacher and student; that is to say, Art stands on a footing of its own, apart from other expressions of man's intelligence. After all, what is the place of the artist in the community? Is he there to discharge merely mechanical functions, such as those of the butcher, the baker, and the candlestick-maker? Is he not rather there to grasp and interpret to his less-favoured brethren aspects of things that only reveal themselves to persons of exceptional faculties in one special direction?

And I need hardly say that I apply this to the Crafts not less than to the Arts. The silversmith, trained in the study of beautiful forms, will find in the materials and methods of his craft motives of design which would never occur to the ordinary mind. He, too, therefore, and within his limits, is an interpreter to others ; he has seized on some aspect of his subject that otherwise would never have seen the light. From this point of view, then, as well as from the others, we see that the career of the artist and the craftsman is not to be lightly undertaken. I suggested to you before, certain obstacles to that career, too deeply seated in the social system to be ignored. I now add to that suggestion the conclusion to be drawn from the consideration of the artist himself, and of the work that he ought to do. And it is that Art must always be a special and exceptional affair, only to be undertaken by those who have a real aptitude for it of eye and hand, and who are endowed with something more than an amiable weakness for the Arts and Crafts. Natural ability, conviction, and enthusiasm are the indispensable basis of your career as artists and craftsmen. I admit that I am putting before you a high ideal, not very easy of attainment, yet I feel sure that for those of you who have that enthusiasm, the picture I have drawn for you to-night will offer no discouragement. Rather, it will appeal to that dogged resolution which is one of the greatest qualities of our race; it will stimulate you to renewed effort to show yourselves worthy of the high calling that you have chosen for yourselves—that of the artist and the craftsman. You in Sheffield are exceptionally favoured; not only have you these admirable schools, not only do you inherit a tradition of that great industry which has made your city famous in the world, and the tradition of that beautiful art of the Sheffield Plating of the eighteenth century, but you have here that alliance and sympathy between the manufacturers and the schools which, as I have already said, I believe to be vital to the right development of the Industrial Arts. I note that in the arrangement of the course of training in the craft of the silversmith and in the award of diplomas, the managers of the School and the members of the Master Silversmiths' Association work together, and I am told that out of some seventy students the Master Silversmiths pay the fees of some thirty students, and allow them time off in order to attend the classes. That is an example of a practice which should be general and it is characteristic of the North in its grip of fact, and of the essential conditions of success. It is to me a clear indication of the path that all our training in the Industrial Arts and Crafts should follow, if it is to be really efficient and of genuine value to the State.

In the remarks I have addressed to you to-night I have referred to the art side of the training given in your school. The technical side of it is rather different. There are not here the same difficulties in the condition of modern society, because the technical trades are concerned with the necessaries of life. We must have machinery; we might, alas, be able to exist without the painter, the sculptor, and the craftsman, but we cannot dispense with the plumber. There must, therefore, be a constant and increasing demand for the technical trades, and here and elsewhere in the North, where people get down to facts with the least possible delay, I may say, as a student of methods of training, that I have nothing but admiration for the way these matters are handled. But here, too, as in the Arts, I would urge you students to be sure that you have an aptitude for the trade you select, and then to do your best to master it. It has been pointed out again and again that what is needed nowadays by employers is not a smattering of general education, not an amiable interest in art, but specialized knowledge and specialized skill. I have heard of cases in which a student of a special craft has been tempted by some minor success to forsake his craft and branch out into more ambitious art. That, believe me, in ninety-nine cases out of a hundred is a mistake. It is not the man who can do what many others can also do that is wanted, but the man who can do one particular thing, if not better than anybody else, at least as well as it can be done, and that is what you should all aim at. It is not given to every one to reach his Corinth, but if you devote yourselves to the study of your trade, your art, or your craft, in this thorough and determined spirit, I do not doubt that, though you may have to make a fight for it, you will find your place in the scheme of life, you will realize yourselves in faithfully contributing your share of work to the life of the community.

Feb. 1913.

FAMOUS MEN [1]

IT is usual for our Gold Medallists to make an Address on such occasions as this. But before I do so, let me thank you, Lord Plymouth, for the extremely kind things you have said—much too flattering, I am afraid, as they often are on these occasions, and also for the graceful compliment you have paid the Institute by coming here to-night to present the Medal : I thank you also, my colleagues, most sincerely for the honour you have conferred upon me. There are honours that may seem to result from a fortunate combination of circumstances, and the recipient may feel like a man who has suddenly come into a fortune, but he does not value them so much as those which he owes directly to the choice of his colleagues : because it is by their judgement in the long run that he stands or falls. They know his limitations as well as his powers ; and if with this knowledge they still feel able to nominate him for such an honour as the Gold Medal conferred by his gracious Majesty the King, he has some ground of hoping that his success is not a mere flash in the pan. I need hardly say that I esteem it a very high honour to be included in the list of our Royal Gold Medallists. There can be no greater encouragement to any architect who still has his eye fixed on the future. But these things lie in the lap of the gods ; and it is well to look backwards as well as forwards, and to endeavour to place ourselves in

[1] An address given at the Royal Institute of British Architects on the occasion of the presentation of the Royal Gold Medal, June 1913.

touch with the mighty men of old. I am a firm believer in tradition. In the pride of youth one is tempted to say, with Sthenelus, son of Capaneus, 'We hold ourselves to be better men than our fathers.'[1]

Yet our fathers before us put up a good fight for what they believed was right, and though the methods and occasion of fighting vary with every age, the essential thing is to remember and maintain that gallant spirit, that high standard of honour, that brave endeavour after noble aims, which are of more value than any particular success. Therefore this evening I shall take as my text the words of the Preacher : ' Let us now praise famous men . . . leaders of the people by their counsels . . . wise and eloquent in their instructions.'

It is a far cry back to that little meeting at the Thatched House Tavern in the year 1834 when some half-dozen architects met together to consider the formation of an Institute of Architects. There were present, among others, Barry, Bellamy, Decimus Burton, Fowler, Goldicutt, Gwilt, and Hardwick ; and of these we may say with the son of Sirach : ' There be of them that have left a name behind them that their praises might be reported . . . and some there be which have no memorial, but these were merciful men whose righteousness hath not been forgotten.' Their buildings have been less fortunate ; so we

---

[1] Ἡμεῖς τοι πατέρων μέγ᾽ ἀμείνονες εὐχόμεθ᾽ εἶναι
. . . .
Τῷ μή μοι πατέρας ποθ᾽ ὁμοίῃ ἔνθεο τιμῇ.
Il. iv. 405–10.

may leave them there, and pass on to Decimus Burton, who, after long years of neglect and oblivion during the days of the Gothic revival, has now come into his own again, and recovered the appreciation that he fully deserved, for he was a very accomplished architect, learned in his art and fastidious in his taste. Few better things in their way have been done in London in the last hundred years than the screen at Hyde Park Corner and the hall and staircase of the Athenaeum. Burton had caught something of the spirit of the Ionian Greeks, of those highly cultivated artists who adorned the cities on the eastern coast of the Aegean. His work is genuine Classic, but it is the Classic of a civilization not so remote as that which inspired the Parthenon, but in a way familiar to us and relatively scarcely less advanced than our own. Burton lived to a great age ; he was not a Gold Medallist, or a member of the R.A., and, though his career must have been singularly successful, when he died at St. Leonards a few years back he was almost forgotten by the general public.

Of the others who met at the Thatched House in 1834 Barry became Sir Charles Barry, Gwilt wrote his immense *Encyclopaedia*, and Hardwick was the well-known architect of Euston Station and of the Goldsmiths' Hall. The Institute was established the same year as this meeting. Lord de Grey was elected President, Donaldson and Goldicutt Hon. Secretaries, and among the Council were Barry, Decimus Burton, Basevi, and Philip Hardwick. Sir John Soane made the new Institute a handsome donation, and in 1837 a Royal

Charter was granted by William IV. All these things are stated in our Kalendar, but I make no apology for introducing them to-night to those of our audience who are not members of this Institute, or even for reminding those who are, of the long and distinguished tradition of the Body to which they belong. It is a good thing now and again to hark back to the hill on which we were born.

I now come to the Royal Gold Medallists of the Institute, and here I have a curious piece of information unearthed for me by our Librarian, Mr. Dircks, to whom I am indebted for some very interesting notes which he has been good enough to collect for me out of the Records of the Institute.

In the year 1846 Queen Victoria consented to grant annually a Gold Medal for promoting the purposes of the Institute, and the Council decided that this should be offered annually for ' designs calculated to promote the study of Grecian, Roman, and Italian architecture '. (You will note in passing that the Council, so far, was faithful to the tradition of classical design ; the possibility of Gothic was not even thought of.) Tite, Charles Barry the elder, Angell, Donaldson, and Sidney Smirke drew up the conditions, and the subject set was ' a building suitable for the purposes of the Institute, at a cost not to exceed twenty thousand pounds '. The result was disappointing. The assessors reported that ' not more than one of the designs possessing the slightest pretension to consideration as an architectural composition could be properly executed for less than double the sum specified '. Our grandfathers did not beat about the bush, and there is a fine flavour of the polemic of the previous century in this extremely blunt announcement.

No award was made, and the Council thereupon revised their arrangements and decided to award the Medal on the basis that holds to this day, for distinguished services to architecture without regard to nationality. It would be impossible to deal with all the names of its recipients. They include famous architects and writers on architecture from France, Germany, Austria, Italy, Holland, and America, in addition to most of the best-known architects of this country during the past three generations. I find that it has been awarded in France to such men as Hittorff, Viollet-le-Duc, the Marquis de Vogüé, Garnier, Choisy, and Daumet ; in Germany to Schliemann and Dorpfeld ; in Italy to Canina and Lanciani ; in Austria to Von Ferstel and Hansen ; in Holland to Cuypers ; in America to Hunt and McKim ; and if you pass in review the names of the Gold Medallists of this country you will get a pretty clear insight into the movement of architecture and the trend of artistic thought from the period when the Medal was established down to the present day. The old Guard was gradually worn down; Cockerell, Barry, Smirke, and Hardwick were succeeded by the champions of the Gothic Revival, and now their day is past and their lesson learnt, and we move again, at least I personally hope so, in the calmer waters of the older tradition, developed and extended by its applications to modern needs. I can select only a few typical names from among the distinguished men who have been awarded the Gold Medal of the Institute.

Early in the list appears the name of Thomas Donaldson, who received the Gold Medal in 1851, and was President in 1863 and 1864. Though not the first to receive the Medal, he did so much for the Institute that we look on him to a great extent as one of its founders. Donaldson was typical of men whom we have always been fortunate in possessing as members of this Society. He was not a great architect, but he was a man of much energy and business capacity, with a high sense of public duty, and he devoted his considerable powers as an organizer and administrator to the formation and development of this Institute. He laid the foundation of a tradition of public utility and high educational purpose which I am glad to say has never been forgotten or abandoned within these walls. He added largely to our splendid architectural library, both in the way of books and drawings, and the badge of office which I have the honour to wear was presented by him to the Institute. Romance appears but rarely in the careers of modern architects, and some, at any rate, of these eminent men had a more adventurous youth than is given to most of us nowadays. Donaldson, who died at the age of ninety in 1885, had gone out to the Cape of Good Hope in 1809 intending to enter a merchant's office; but he joined a force of volunteers that was proceeding to the attack on the island of Mauritius in the hope of obtaining a commission in the Army. As, however, the French retired without firing a shot,

Donaldson's vision of military glory vanished. He returned to England, entered the school of the Royal Academy, travelled widely in Greece and Italy, became an architect and Professor of Architecture at University College, and devoted a long and most useful life to the public and professional aspects of architecture, and to the development of research into all that concerned the history of the art.

Charles Cockerell, who received the first Gold Medal in 1848, was a few years older than Donaldson, and represents, to me at any rate, the other type of architect—the man absolutely immersed in his art, a scholar and an artist with a passionate enthusiasm for all that bore on the history and technique of architecture. That enthusiasm never flagged to the end of a long and fortunate life. I have heard Norman Shaw describe the fascination of the lectures that Cockerell gave at the R.A. when he himself was a student there. Whatever his subject, Cockerell was very soon back among the scenes of his travels and adventures. He forgot his audience in living again those brilliant enterprises of his younger days; and went on pouring out reminiscence after reminiscence till something recalled his attention to the fact that he was not in Greece or Asia Minor, but in the Lecture Room of the Royal Academy. Cockerell—who, besides being a beautiful draughtsman and a sensitive artist, was a fastidious gentleman—had certainly exceptional advantages, but he used them well. He steeped himself in the architecture of Ancient Greece, and carried into his own work something of its delicate and austere reserve. That an artist of such enthusiasm should have his limitations was inevitable. A certain coldness of temperament and a certain academical perfection and propriety may sometimes arouse in more warm-blooded artists an irresistible desire to kick over the traces ; but his buildings have always a distinction rare in modern architecture, a certain well-bred personal quality that reveals itself as something beyond the reach of merely conventional accomplishment.

Sir Charles Barry received the Medal in 1850, and on the death of Lord de Grey, who had been President of the Institute from 1835 to 1859, he was offered the Presidency, but declined it, probably for reasons of health, for he died in the following year. Barry was a thoroughly well-trained architect, and it is to be noted in the case of nearly all these famous men that they devoted a good deal more time both to their apprenticeship and to subsequent study abroad than is the fashion at the present day. Five years' apprenticeship, followed by two or three years' study of ancient buildings abroad, was by no means unusual in the training of architects eighty years ago ; and though fashions change and the technical detail of that generation may be out of favour with this, there can be no doubt that these men were thoroughly well trained in the technique of architectural design, the more so as they were able to concentrate on it exclusively, instead of having to devote a considerable part of their energies to the acquisition of that applied science which has become a necessary part of the equipment of the modern architect. Barry travelled extensively in France, Greece, Turkey, Syria, Palestine, and Egypt, and this Institute is fortunate in possessing the diaries of J. L. Wolfe, his travelling companion during these three years. Quite recently a very high compliment was paid to Barry in these rooms by a well-known American architect. Mr. Hastings referred to him as one of the most remarkable architects of the nineteenth century, for his powers of planning a big design. Some of his detail is out of fashion and rather dull, but his great ability as an architect is so generally recognized that I need not remind you of his buildings. Two points, however, are noticeable in his work : signs of the rift in the great tradition of English Classic, warnings of the upheaval that was to supersede it. The first is his choice of model, the second his complete surrender of it on a memorable occasion. Whereas Cockerell had definitely elected for Greek models and inspiration, Barry reverted to the more florid traditions of the Italian Renaissance, even following Italian originals pretty closely in his designs for such clubs as the Travellers' and the Reform. Up till comparatively recently Barry's lead was followed in most of our public buildings. Now the pendulum has swung back to Greek motives seen through French spectacles. My personal impression is that both Cockerell and Barry were a little off the line, and that those who have blindly followed either the one or the other of these distinguished men may perpetuate a fundamental mistake, that of a too direct revivalism and reproduction, which must be sterile in its results however ably it is done. Had either of these men picked up the simple tradition of English Classic at the end of the eighteenth century, and used it frankly to meet the conditions of the day, we should have been spared years of wasted effort ; but owing to causes far too intricate to be touched on now, the Lord of Misrule had flung his cap into the arena of architecture, and the first momentous intimation of this was the decision, forced upon Barry, to design the Houses of Parliament in the Gothic manner. There is a suggestive sentence in the Report of the R.I.B.A. Council for 1839. Referring to the Commission appointed to investigate the stones to be used in building the Houses of Parliament, it says : ' The investigation may lead perhaps to the adoption of a stone [1] more brilliant in hue than those at present in general use, so as to shed somewhat of the glow of an Attic or a Roman tint upon the architectural features of the public edifices of London ': a pious aspiration scarcely realized in the Houses of Parliament designed by Barry with details by A. W. Pugin. There is no need to revive the worn-out controversy as to who did it. Probably it was a genuine case of co-operation, Barry giving the scheme and general arrangement, and Pugin the detail—detail, by the way, as good as anything of its kind that has ever been done in modern Gothic.

Pugin never had our Gold Medal ; in the light of what followed he surely deserved it, for it was the zeal and enthusiasm of this apostle of modern medievalism that brought out the fighting qualities

[1] The stone actually used was a very unfortunate choice. It is already failing, and the renewal of the pinnacles, crockets, and other embellishments in the Gothic manner is a constant source of expense.

of the younger generation, and won the day for Neo-Gothic. When one considers that there were solid men such as the Smirkes, the elder Hardwick, and Tite, who practised their weighty Classic with unvarying success, it was a remarkable thing to have done. Later on, Tite, who became Member of Parliament for Bath, made a violent attack on Scott's Gothic design for the new Government buildings and, faithful to his convictions, founded the Tite Prize of the R.I.B.A. for the best design of a given subject, according to the methods of Palladio, Vignola, Wren, and Chambers—a counterblast to the Pugin Studentship, established some ten years earlier, for the promotion of the study of the medieval architecture of Great Britain and Ireland.

Hardwick, it is true, designed the Lincoln's Inn Library, but I have always understood that the late J. L. Pearson, R.A., was a young man in his office at the time ; and Hardwick's real quality as a designer is best shown in the Propylaea and the impressive Hall of Euston Station, and in the Goldsmiths' Hall.

Sir Robert Smirke, the architect of the British Museum, takes us back into the eighteenth century, for he was born in 1781. He was made an R.A. in 1811, and received the Gold Medal in 1853. One of the best of his buildings, and one of the best examples of the masculine Classic of his time, the General Post Office, has disappeared within the last year, not without a gallant effort to save it on the part of this Institute. Sidney Smirke, his younger brother, who designed the Reading Room in the British Museum, the Oxford and

Cambridge Club, and the Carlton Club,[1] was awarded the Gold Medal in 1860, and from 1861 to 1868 was Professor of Architecture at the Royal Academy, a post which has now been filled by five of our Gold Medallists. The Smirkes were, I take it, the last representatives of a tradition of Classic derived from Sir Wm. Chambers. Robert Adam's manner, graceful and accomplished as it was, was to some extent an original invention of his own, as indeed he believed it to be himself. Cockerell's manner was not less personal than that of Adam. The final version of Chambers's ideas of civil architecture, somewhat debased and a good deal vulgarized, appeared in the work of Tite and Sidney Smirke.

In this rapid survey I have now come to the point at which we reach men with whom some of us, at any rate, were personally acquainted. We have passed the disastrous days of the great Exhibition. Digby Wyatt, a man of wide knowledge but no definite bent in design, received the Gold Medal in 1866 ; but I take it, it must have been a little in the nature of a consolation prize, for the eclecticism and compromise of his generation were things of the past, architecture was deep in the whirlpool of the Gothic Revival, and the cry was raised, that is being raised again to-day, that the architect with his T-square is the source of

[1] In the Carlton Club Smirke reproduced Sansovino's design of the Libreria Vecchia at Venice in Caen stone with Aberdeen granite columns. The Caen stone weathered so badly that in 1923 it was found necessary to reface the whole of the Club in Portland stone, and this was carried out to an entirely new design by me, 1923-4, the old door and window-openings being preserved.

all evil, and that salvation is only to be found in the untrammelled genius of the working man. But the architects were energetic and astute, and they rode the storm with remarkable skill.

George Gilbert Scott, who received the Gold Medal in 1859, was President of this Institute from 1873-6, and was, I take it, quite one of the ablest men of his time.

How many hundreds of churches he dealt with has never been known, possibly Scott never knew himself. There is a story that I had from a well-known pupil of his, that Scott once found himself at a remote station in Yorkshire, and was compelled to wire to his head clerk : ' Why am I here ? ' Probably no other architect has ever left his mark on the historical buildings of his country to such an extent as the late Sir Gilbert Scott. In his *Recollections*, written in 1873, he stated : ' I had been one of the leading actors in the greatest architectural movement which has occurred since the Classic Renaissance.' The value of the movement is open to question, but there can be no doubt of the fact that Scott was for a time its most redoubtable protagonist ; and the *Dictionary of National Biography* informs us that ' his excessive energy in restoration and renovation led to the formation, in the last year of his life, of the Society for the Protection of Ancient Buildings '. I fear our generation is not particularly grateful to the zeal and enthusiasm, amateur or professional, of the 'sixties and 'seventies. From the point of view of professional practice those days must have been a glorious time for architects. There were not too many architects about, the landed interest was extremely prosperous and ready to support its views on ecclesiastical art by putting its hand deep in its pocket. Everywhere there was a fine glow of sentiment and romance, unimpeded by a too exact knowledge of the facts of architecture or practical understanding of its functions. A heavy reckoning has had to be paid for those happy days of romance. It is not only that our historical buildings have suffered. That has happened elsewhere, as in France, to an even more disastrous extent. The real mischief has been the confusion that has arisen between architecture and craftsmanship—a confusion that eighty years ago would have been inconceivable—and the result of this ill-balanced zeal for craftsmanship was that the purpose of architecture was all but forgotten in England, and it is only within the last few years that there has appeared unmistakable evidence of a return to a saner tradition. It is useless to write history backwards, but one cannot help speculating what men of such undoubted ability as George Gilbert Scott, Street, Pearson, or Bodley might have done for modern architecture if they had been trained in Classic design instead of in the details of Gothic.

Yet as the movement approached its end the conviction of its leaders became almost fanatical. In 1855 Street had written : ' I have no reason whatever for doubting that if we wish for a purer school of art we must either entirely forget the works of the Italian Renaissance architects, or remember them only to take warning by their faults and failures.' Some twenty years later Street could hardly forgive Bodley for straying beyond the orthodox boundaries of Gothic into the amiable French Renaissance of the London School Board Offices; and he himself nailed his colours to the mast in the last great effort of his life, the new Law Courts, a really monumental work, however much one may criticize it in detail. Street was not only an able architect ; Norman Shaw used to say that Street was a man who would have made his mark in any calling that he had put his hand to, and, though without academical training, he wrote excellent English. He was also a man of strong convictions, and a very dominant individuality. My impression of him remains as I saw him in 1880–1. I was working against time in the schools of the Royal Academy, being indeed anxious to get away for a cricket match in the country ; our old friend, Phenè Spiers, brought in a burly, bearded man, who tramped across the room and asked me what I was doing. In my haste I answered shortly, but was met by a good-humoured smile, and the visitor retired. I learnt afterwards that this was Mr. Street, and the impression that I formed of him as a strenuous and most capable personality, strong in his views, and indifferent to convention, was I believe the right one. I just recollect, too, that memorable election, in the last year of his life, when the forces of Art and those of Business were set in battle array, and Art won a brilliant victory : a victory cut short, alas ! by Street's untimely death.

Since those days we have learnt from adversity the necessity of combining business aptitude and art. Since those days, too, the battle of the styles has dropped into oblivion. The point of view has shifted, or rather we have come to see that all vital art must be a personal expression—that architecture, not less than the other arts, is the expression of an idea, with this condition added, that it must also be the fulfilment of a particular and specific need. Thus these questions of archaeology fall away of themselves. We use in architecture a language based on the past, just as in common parlance we use the language which has resulted from long generations of use ; but we do not use language for the sake of using it, we use it to express a definite idea, we have no more use for the mere stylist than we have for the mere rhetorician. The days of the revivalist are, I hope, finally numbered.

But I have wandered from my point. I set out to praise the mighty men before us, and on that note I should like to conclude my Address. We live so fast nowadays that we have little time to look behind us ; yet it is well to pause now and then to pick up our place in the line of long descent, and to remember the tradition of the past. This Institute has been in existence for nearly eighty years. It is second in point of age only to the Royal Academy and the Royal Society of Painters in Water Colours. I have mentioned to-night a few only of those who in past years have played a great part within the walls of this Institute. Others, scarcely less distinguished, might well be mentioned, and I have said nothing of our contemporaries. Yet I have hoped to suggest to you something of the great tradition of this Institute, and to recall to your memory the part that it has played in the development of modern architecture. I do not doubt that that tradition will be worthily maintained by this and succeeding generations. We ourselves are in the position of trustees for the younger generation, and we are bound to take a far-reaching view of the duties of our trust. Much of the work of the Institute must necessarily be concerned with details of administration, and members have always given their services for the purpose in the most ungrudging spirit. But a wide outlook in the arts is in accordance with our best tradition, nor do I think its members are likely to forget the high purpose for which this Institute exists, for the advancement of architecture, ' usui civium, decori urbium '.

June 1913.

THE OUTLOOK OF ARCHITECTURE (1913)[1]

IN an address which I had the honour of giving in this room on a recent occasion I reviewed in a very cursory way the ups and downs of architecture in this country during the last hundred years. I brought my survey down to the rise and gradual failure of the Gothic Revival. To complete the main outlines of the picture, it is only necessary to remind you of the reappearance of Classic, and its gradual consolidation within the last twenty years. At the moment of its triumph, Neo-Gothic was already undermined by the rhetoric of its advocates, and even by the adventurous spirit of some who had been trained in the strictest sect of the Pharisees. It is now many years since raiding expeditions into the territory of the Renaissance were made by Devey, Nesfield, and Norman Shaw ; and these have been followed up by a systematic study of Classical architecture which has resulted in the recapture of some at least of the scholarship of the art. Undoubtedly interest in architecture is more widespread than it was, and our literary friends are well to the front, telling us of our failures, what we ought to do and how to do it. Experienced architects are not very likely to be turned from the course they have set

[1] The opening Address of the Seventy-ninth Session of the R.I.B.A., 1913-14. Delivered at the First General Meeting, Monday, 3rd November 1913.

themselves by criticism and clamour, but the rising generation may feel some doubt and perplexity, and I think the time has come to take stock of the situation so far as it is possible to do so.

It is not an easy thing to do, and I must ask for your tolerance if I seem to you to misread the signs of the times. It is difficult to appreciate contemporary art with any certainty. One cannot get far enough back from it to place its features in right perspective. The tendencies that result in history do not lie on the surface, and what appears to be a new light may be only the will-o'-the-wisp of a passing fashion. Moreover, the problem of architecture is very complex ; and as the power of literary expression is seldom in ratio to technical knowledge and ability, our guides and critics may possibly misapprehend the situation, and leave unnoticed those strong impulses in artists themselves which must be the foundation of any real progress in the future. Our critics do not always grasp the continuity of architecture, and its solid basis in facts, and I believe it is this omission which explains their hankerings after new styles and their clamour for originality, no matter whether it is good, bad, or indifferent. For some generations art criticism has suffered from a certain feverish impatience, which has blinded it not only to the intimate and necessary connexion of the architecture of to-day with that of the past, but also to the germs of future development, latent in that contemporary art which it is the common practice to minimize and disparage. It is only a few years back since critics, whose training should have given them more insight, complained of a lack of initiative in those who through good report and evil steadily pursued our national tradition of Classic architecture. Time has justified those men, and a very few years have shown the practical certainty of disaster that waits upon jumps into space.

That point of view has been dropped by serious critics, and our professional writers are too well informed to believe in the value or even possibility of any violent cataclysm in architecture, such as that now being attempted by the Futurists and the Cubists in painting and sculpture. We, at any rate, know that architecture is too serious an art to pay any attention to quack remedies. Meanwhile, architecture, or perhaps I should say architects, are attacked from another quarter, and the attack, I do not know whether consciously or not, is a repetition of the polemic of the Neo-Gothic enthusiasts of the 'seventies. A clever writer in the *Morning Post* has drawn a charming picture of those glorious days when Gothic architecture was run entirely by the Guilds : when the workman, unchecked by the architect and his T-square, was working his own sweet will as a free and glorious artist ; when the building craft was the greatest in the world, and the Guilds were its embodiment, storehouses of knowledge, ' the vat ', if I may quote his words, ' into which the experience of all ran '. The master masons, he asserts, were ' cultured men, the associates of Princes and Scholars ; they built with extraordinary audacity and imaginative resources '. We are now told that architecture has lost this fount of inspiration, and we are bidden to throw over our scholarship, our draughtsmanship, our powers of design, our trained technical ability, and watch the 'felicity of action and latent understanding' with which 'a mason tosses and turns a brick'. (I may mention in passing that the Bricklayers' Union would very soon be on his track if he did !) The writer, Mr. March Phillipps, is so haunted by the idea of an architect that he goes so far as to say that he never met a man, other than an architect, who had a good word to say for the architecture which ranges from the reign of James I to that of George IV. I think he must have forgotten Greenwich Hospital and Hampton Court, St. Paul's Cathedral and Somerset House; and without desiring to enter into controversy one is compelled to question the historical accuracy of Mr. Phillipps's charming idyll. I seem to detect the trace of an ingenious theory which a few years ago was spun round an obscure association known as the Comacine masters. Were the Guilds the last refuge of the building art ? the high-minded guardians of all that was noble and beautiful in architecture, thrust out of place by an arrogant intellectualism ? Were they not in fact so hopelessly corrupt in their latter days that the 'adverse legislation', as Mr. Phillipps calls it, became an absolute necessity of intelligent government ? I would ask also, were the medieval workmen the consummate masters of the building art that our critics would have us believe ? Were the master masons the associates of princes and scholars 600 years ago any more than they are to-day ? Is it not an historical fact that many of them built extremely badly, that church towers of the fifteenth century have simply collapsed in France, that some of their most ambitious ventures in construction, as at Beauvais, failed almost at once, and had to be precariously maintained by a network of iron bars? Those who have had the handling of old buildings have had it driven into them again and again that the average building of the Middle Ages was inferior rather than not. I am talking simply of building, not in any way of design and details of ornament, and I say deliberately that at the beginning of the sixteenth century most of the master builders were bad builders; and if our critic has any doubt on the matter, I would remind him of what happened in France in the reign of François I, and of the contemporary evidence of Philibert de l'Orme on the master builders of his time.

Our critic imagines a divorce between the modern architect and his workmen that does not exist. 'Labour', by which is meant the skilled labour of the building trades, is not in 'the state of helpless ineptitude and dull impotence' which Mr. Phillipps supposes. The architect is not a truculent and arbitrary tyrant, any more than the workman is a heaven-born but down trodden artist. They are both, let us say, honest men trying to do their allotted work, and some of them do it exceedingly well. All good architects value a good workman; the unsympathetic attitude of architects is wholly imaginary, and the phrase 'the untravelled workman' which Mr. Phillipps imputes to me, was, if I recollect aright, the invention of a somewhat intemperate champion of the Art master, and I am not conscious of ever having used it at all. Mr. Phillipps makes a distinction between 'creative construction' and 'imitative construction'; the first he identifies with Gothic architecture, the second with Classic. Surely this begs the whole question; this distinction, which is to be the key to the architecture of the future, is only a repetition of the outcries of Ruskin. Nobody, no practising artist at any rate, ever thought about such things before his time. The idea is of purely literary origin, it has no justification in history; on the contrary, it makes the serious error of overlooking the work of tradition in both medieval and Classical architecture, that slow and almost unconscious moulding of architectural forms that keeps on slowly moving from generation to generation. It is an idea that has arisen from the habit of regarding the details of architecture as architecture itself, of concentrating attention on words rather than on language. Nor, as a student of the history of architecture, is one in the least disposed to accept the assertion that the architect is the *fons et origo malorum* in architecture and that he is so by reason of his trained ability, for that is what the charge amounts to. The more closely one studies certain contemporary criticism of the arts, the more convinced one is that it is inspired by the dictum of the celebrated 'Capability Brown', that 'knowledge hampers originality'. Mr. Phillipps says that in medieval building there was not 'a sign of a dictate, automatically delivered and passively accepted', but he has himself to admit in more 'important operations' the work would be 'supervised by some craftsman of more than local repute'. Indeed, unless human nature was different in kind in medieval times from what it has been both before and since, building operations could only have resulted in Towers of Babel unless there was somebody in control whose dictates were both delivered and accepted. That he was not equipped as a modern architect we are all agreed, but that he was a person of superior knowledge in control of the workmen is also certain—it is immaterial what he was called; and this knocks on the head the engaging theory of the workman and his own sweet will.[1] We have to get back to the facts, and I have dealt at some length with this criticism of modern architecture because Mr. March Phillipps writes so well that some danger to the right understanding of the art lurks in his well-turned sentences. The views that he advances are, I think, off the track of history. No serious advance is to be made by turning our back on the immediate past, or blinking the facts of the present and trying to jump the centuries. This idea that the hope of architecture lies in the untrammelled (not untravelled) genius of the British workman is the merest *ignis fatuus*. Any one who has first-hand acquaintance with the condition of modern building, with the methods of modern construction, with the qualifications and habits of the modern workman, knows that the suggestions of our critics are impossible in practice, and that even

---

[1] In the Opere del Duomo at Sienna there are five well-executed detail working drawings of the Cathedral, including a design for the Campanile (octagon on square) by Lando di Pietro (1339). The drawing measures 7 × 1 feet, and this and the other drawings are as carefully and accurately detailed as a working drawing by a modern architect.

if they were possible the result would probably be an exaggerated version of the efforts of the speculative builder. The man of genius who first made popular this delightful dream of medieval art had the excellent sense to call his message ' News from Nowhere '. Morris's theory of architecture was just the expression of his own temperament, and the logical corollary of his political views and of his personal conception of architecture as the drudge and vehicle of decoration on the one hand, and of practical necessity on the other. It is a new view constantly reappearing in modern criticism, but I would remind our critics that architecture is the greatest of the plastic arts, and that it is not its function to sit at the feet either of the ornamentalist or of the engineer. I do not think that architects are seriously alarmed as to the future of their art. They will agree heartily with Mr. Phillipps in his search for simplicity and sincerity of statement. Where they will entirely decline to follow him is in his subordination of architecture to the ignorance and incompetence of ' average labour '—(the phrase is Mr. Phillipps's, not mine), and I may add that under present conditions labour is likely to make an end not merely of architecture, but of any building at all (1924).

So far I have endeavoured to put before you what I may call external criticisms of architecture. We cannot entirely disregard them, because they are widely read by the general public, and as they are usually stated in readable English they may have a far-reaching and unfortunate influence, against which we have to be constantly on our guard. The difficulties in which the art was landed by the unbalanced eloquence of a great writer in the last century are a matter of common knowledge.

Now let us consider the art from our own point of view. Any one who has studied history knows how slow and gradual has been the growth of architecture, by centuries in medieval times, by half-centuries from the dawn of the Renaissance down to the end of the eighteenth century. These advances, too, have been made not by deliberate intention, but almost on compulsion, in order to meet the changing needs of a constantly expanding civilization. Looking back on the past we can trace the successive steps, we can show the development of construction and the gradual perfecting of technique, and we can follow more obscurely the trend of artistic thought, the gradual consolidation of those impulses which lie at the back of vital movements in the arts. The road is unbroken—where we miss it, there is no hiatus in fact, but only in our knowledge of the facts, and if there is one thing more certain in history than another, it is that of all the arts architecture is the most steady and consistent mover. The idea of the Futurists that architecture will advance by being turned upside down is not worth the consideration of serious students.

On the other hand, the arts do not stand still, architecture least of all, because it is essentially a practical art. Fresh problems present themselves in planning, provision has to be made for the ever-widening range of applied mechanical science, new methods of construction have to be considered, the practice of architecture becomes more difficult every year, and the modern architect has to deal with a range of subjects which would have paralysed his grandfather. The question we have to consider is how far these changed conditions are likely to affect design, and how we architects should set our course if we do not wish to drift on to the quicksands of futile experiment. It has sometimes been suggested that the future of architecture lies in a resolute rejection of all the accepted forms of architectural expression. In Barcelona architects appear to be following the precedent of cavemen. What we are to do after this is not quite clear, because some of our critics tell us that we should leave our steelwork and our reinforced concrete just as it is, and others want us to spin new forms out of our inner consciousness. Our critics are so dreadfully impatient. Architects may well say, like the unfortunate debtor, ' Have patience with me and I will pay thee all.' But that is just what we are not allowed to do, because our public is never quite sure whether we are the enemy of society, or the *Deus ex machina* who can resolve every conceivable difficulty. The past fifty years have seen some desperate endeavours to invent something new, experiments in various styles in the past, and experiments in what is fondly believed to have no relation to the past. I think it is time we gave up these conscious and artificial attempts at originality, and let it find itself. Where our critics go wrong is in demanding a new language when they ought to be demanding new ideas. The old language will do very well if we are masters of it and have the brains to use it to the full.

Meanwhile history has been making itself, and

making itself in a rather curious way. If we go back to the last quarter of the nineteenth century, we find that the orthodox Classic of the older school had dwindled away to dullness and what for want of a better word I have to call stodginess. The Gothic revivalists had broken loose in all directions and afterwards lost their clue, having condensed into some thirty years all the variations of an art that had taken five centuries to run its course. The more original among these men had for years been feeling their way out with tentative excursions into the Renaissance : Nesfield at Kinmel, Devey in many a picturesque country house, Norman Shaw, who with all his genius in design reached his Classic too late in his career, and close on the heels of these came men who I am glad to say are still with us, and who won their spurs when some of us were still in our articles. Meanwhile a generation has grown up no longer content with odds and ends of detail, however picturesque, but anxious to get to the heart of things, and to grasp the informing spirit of Neo-Classic architecture. The technique of the art in its widest sense, not only in the nuances of detail, but in the larger aspects of planning and composition, rhythm, and proportion, has received in recent years a study and attention such as had not been given to it since the days of Cockerell, and we have now before us versions of Neo-Classic which deserve to be taken seriously, and out of which, I believe, may ultimately develop that standard manner which is essential to the appearance of any such vernacular art as existed in civilized Europe in the first half of the eighteenth century. I think all close observers of modern architecture will admit this real advance, and this gradual *rapprochement*, as I have to call it, of the ablest designers that we have. It exists so far in a common point of view rather than in an identical manner ; because we have varying versions of Classic all worth taking seriously—the attempt to pick up the thread of Cockerell's tradition—a somewhat dangerous leaning towards the fashion of our colleagues in France—and the more sober manner based on our own Classic of the earlier part of the eighteenth century. That any one of these should sweep the field entirely is neither to be expected nor to be desired. Such a result would be alien to the genius of our race for individualism, and its robust dislike of pedantry.

Nor would it be a complete synthesis of all the factors in the case, for ecclesiastical architecture has yet to be taken into account. Our English clergy still cling to Pugin's totally unhistorical claim that Gothic is the only possible form of religious architecture, and, Classical churches being ruled out of court, our architects have to persevere with Neo-Gothic. Let me say at once that some of them design in it with great ability, and that, so far as my observation goes, the architects of this country are the only ones who have got within range of the subtle and elusive spirit of medieval art, so far as it is possible for any one to do so. Then, too, there is that Byzantine strain which found such masterly expression in Bentley's church at Westminster. Its influence is less marked than it was, but it has been a valuable factor in the advance of architecture, because in its austerity and reserve, in its feeling for surface ornament and the value of abstract form, it is akin in spirit to the purer forms of Classic art. All these elements the wise artist has to note, and, in spite of their different idioms, he may find a certain bond of kinship in their constant effort after simplicity of statement, and even the most ardent classicist may learn a lesson from the elasticity and resourcefulness of Gothic.

Let me say at once that I am not advocating the eclecticism that has done duty for design in the past. Every artist has to find his own personal method of expression, but the wider and deeper his range of study, the more flexible and the more assured will be his art. Craftsmanship, in the sense of the dexterity of hand acquired by specialized work in one direction and on one material, is an admirable thing, but it is not architecture, nor does it represent the aim and ideal of an architect in regard to his art.

The only effective source of development in architectural form must be new conditions of building, and this will be very different from that new and original style for which our critics hanker. The fashion of ornament may change, but the problem of architecture does not lie with ornament, and the epoch-making discoveries in the art have arisen from practical necessities handled in the most direct and even uncompromising manner. Witness the Colosseum and the dome of the Pantheon. The designers of these great buildings did not trouble their heads about inventing fresh detail ; what they found to hand was good enough for them. Where the Roman architect was so great, greater even than the Greek, was in the masterly handling of a great conception, in that power of bringing the mind to play on the actual facts. The American skyscraper is a less fortunate example of a new form arising from new necessities, though the solutions have not been happy, because the essential elements of tower design have been forgotten. It comes, I think, to this, that although new architectural forms in the sense of new outlines, new groups and masses will naturally develop out of the changing problems of civilization, no necessity arises for anxious effort to change the ordinary vocabulary of architecture. Perhaps of all futile experiments in originality the competition for a new French order to glorify Louis XIV was the most gratuitous and the most ridiculous.

The question still remains how we are to deal with inventions such as reinforced concrete when used for the exteriors of buildings. Are we to adhere to the shibboleth of the Gothic revival, and show our construction naked and unashamed, and are we to suppose that our aesthetic sense will alter so materially that we shall presently find pleasure in ranges of openings supported and separated by the thinnest piers to which the engineer can reduce our points of support? What may happen to our aesthetic sense in the future no one can say in view of the chronic assaults made on the sanity of the public. The only evidence is what has happened in the past, and that evidence shows that though from time to time there have been eccentric aberrations, the orbit of taste has ranged between fairly determinable points, and those points have not included such skeleton building as is of the essence of steel construction or the brutal texture of crude concrete. If, as I incline to think, our dislike of it springs from some deeper instinct than mere unwillingness to change, we must reserve our freedom to use inventions such as reinforced concrete as mere instruments of building in the same manner as we use steel construction, or as the Romans used their system of brick ribs and arches. I see no reason why, in dealing with this and similar methods, we should not avail ourselves of all the weapons in our armoury, translating our construction into such forms as will best express the central conception of our design. In other words, we are not compelled to subordinate our design to the instruments we employ. There is always a touch of the autocrat about the Mistress Art.

What conclusion is to be drawn from the considerations I have endeavoured to put before you? What is it we should aim at, and how far are we likely to realize our aims?

We have to accept the fact that we are at the end of 150 years of eclecticism. The last genuine tradition died with Chambers. His successors carried on his manner, but other elements had come into play, the Romantic movement on the one hand, and the age of archaeology on the other. Architecture, most unfortunately, came within the literary net, and it has not yet escaped it. Hitherto —that is, till the latter part of the eighteenth century—architects had studied old work assiduously, but it was with the object of perfecting their technique. The archaeologists have worked with quite different objects, and though they have done invaluable work in extending and correcting our knowledge of the past, their labours have had the curious result of placing architecture on the wrong issue, and of reducing architects from time to time to a state bordering on imbecility from the very profusion of the details at their disposal—a fact that will need attention in the conduct of the new British School at Rome. From this state of things I think we are emerging ; the limits and the relation to each other of architecture and archaeology are becoming clearer, and the conviction has been steadily growing in the mind of architects that details are but the outside of the cup and platter, and that their value is conditional on the use that is made of them. This is the first step towards the reorganization of architecture and its recovery from the chaos of the nineteenth century.

We cannot escape the difficulty of modern architecture, that we are offered too wide a choice, that there are too many wells to draw from in the interminable issues of photographs and illustrations, and also that there are too many fashions set by irresponsible people. The temptation to yield should be met if our architecture is to be robust ; and the way to meet it is to shape differently at the problem of design, to search for the idea, and let the form develop out of it. A master idea carries with it its own expression, and, to a mind well stored with the language of architecture, the form follows the idea so closely as to be almost inseparable from it.

I am not going to attempt any prophecy as to

the future of architecture in this country. My own view is that genuine progress is likely to be made only along lines already laid down, by the skilful use of opportunities as they occur in plan and construction, and by the watchful care of all elements in design that pull the same way, namely, in the direction of strength, refinement, and sincerity of statement. Our French colleagues, I am told, deprecate our ventures in monumental Classic, and would urge us to follow the models of Late Gothic or even of Jacobean architecture. I can only suppose that these gentlemen are unacquainted with the work of Wren and Vanbrugh, Hawksmoor, Gibbs, and Chambers. In the work of all these men there is latent a tradition, still unexhausted, still capable of development and application to the problems of modern architecture ; and this question of tradition is of the first importance. We ourselves are, I believe, slowly moving towards the only possible standpoint in gradually concentrating on the tradition of English architecture of the eighteenth century, and our French critics seem to me to have shown little wisdom in deserting the splendid legacy of the Gabriels, father and son.

Our course then is clear. We are not to be rushed by the outcries of our critics, or moved to hurry by frantic attacks on architects, made sometimes for reasons quite unconnected with architecture. We should pursue our steady way, strong in our knowledge of the past and in our faith in the future, and in that enthusiasm which is the privilege of creative artists. For this kinship of artists should be the real bond of union between architects, the source of that honourable fellowship without which individual efforts must too often fail. And in concluding my remarks I would urge the value, and indeed the necessity, of this *esprit de corps*. The career of an architect is by no means an easy one. Unforeseen difficulties may arise in his way, and he may need that helping hand which, I hope, will never be refused by his colleagues. In all such cases we should stand by our brethren. Moreover, there are intricate and difficult questions to be determined by the profession in the near future,[1] one in particular which has blocked the way for a generation and which has for years received the anxious consideration of successive Councils of this Institute. I have every hope that at a near date your Council will be in a position to offer you its considered suggestions for the solution of that question. I will only remind you that such questions can only be settled by pulling together. You will recollect Aesop's fable of the bundle of sticks. It is not to be supposed that any method can be devised which will be wholly acceptable to everybody. But when the solution to which I refer is suggested to you, I feel sure that this *esprit de corps* will have the full scope and bearing that it should have in a great profession such as ours, and that we shall not trust in vain to this same honourable sense of fellowship.

[1] Since settled by the amalgamation of the Society of Architects with the Royal Institute of British Architects, 1924.

ATAVISM IN ART[1]

BY atavism I mean the tendency to throw back to racial instincts, to strains of thought and temperament that would seem to have been long lost beneath the developments of later ages. I may say at once that I shall suggest speculations which it is impossible to prove or disprove. Yet History is something more than a catalogue of facts, and here and there we come upon lacunae which can only be bridged over by an effort of imagination, and for which some sort of working hypothesis is necessary, if our data are to have any meaning for us at all.

The origin of Gothic architecture is one of these lacunae. How are we to account for its sudden leap into brilliant life in the latter part of the twelfth century? Why is it that it should have done so at that time in the west of Europe, in France that is, and in a less degree in England, rather than in Central, Southern, or Eastern Europe? We may or may not accept the technical explanation of its development given by writers such as Viollet-le-Duc, who built up out of the architecture of a certain date and area in France a system of scientific construction which was certainly never in the consciousness of medieval builders. The pointed arch followed close on the heels of the round arch, and though it may have been little more than an accident in the first

[1] Read at the British Historical Conference 1913, and to the Oxford Archaeological Society 1916.

instance, it has been shown again and again what amazing developments followed on this casual motive, how the art of vaulting grew, how it in turn was modified by the beautiful art of the glass painter, till the solid walls of early pointed became a mere framework for the windows, with a complicated, and sometimes inadequate, system of buttress and counterweight to hold them up. All these explanations are a valuable commentary on what is left of this great period, but they do not exhaust the subject. There are even awkward obstacles in the way of the theory that Gothic architecture was merely logical development from Romanesque. The pointed arch existed in Syria before the pointed arch of Gothic, yet it led to no such architectural developments as we find in Western Gothic. The domed churches of Angoulême and Périgueux show that the pointed arch could be used in a style alien to the method of Western Gothic. Lastly, there is the plain fact visible to the eye of any observant student, that the spirit that inspired Romanesque, even when the immature art of barbarous peoples is discounted, was different from that which lay at the back of Gothic. Its evident feeling for massive strength and simplicity of statement, its vague reminiscence of an older system of proportion and spacing, show that Romanesque architecture was the last thin echo of the classical tradition, rather than the voice before the dawn of medieval art.

Suddenly, in the last fifty years of the twelfth century, a change in direction appears, not very marked at first, but gathering volume as it goes

with astonishing rapidity. Some fresh element has
come into play, some instinct, dormant hitherto,
has sprung into active life, and found its expression in methods of architecture which, at any rate
in their complete development, were different in
kind from all hitherto known architecture in the
Western world. It is an astonishing fact that nearly
all the great French cathedrals were built within
the hundred years from 1150 to 1240. When
Philip Augustus died in 1223, the cathedrals of
Notre-Dame, Chartres, Bourges, Noyon, Laon,
Soissons, Rouen, Évreux, Bayeux, Coutances,
Mans, Angers, Poitiers, and Tours were in the
main completed. One finds much the same state
of things in England, and within one or two
generations later the change had spread over
Western and North-western Europe. Without
going farther into details we may take the general
fact that about the end of the twelfth century
there came about this extraordinary development
in church architecture which had its counterpart
in the castle, the house, the Town Hall, and the
Palais de Justice. What was behind it all? Viollet-
le-Duc, most clever and plausible of archaeologists, explained the rise of Gothic as due to
impatience with the feudal system. The bishops,
he says, put themselves at the head of a popular
movement, and built their cathedrals as ' une protestation éclatante contre la féodalité ', and elsewhere he asserts that the architecture of the
beginning of the thirteenth century was the
' purest and most exact reflection of the ideas of
the nation at that epoch '. This was a generaliza-
tion such as the French archaeologist loved ; but

it hardly covers the facts. If the cathedrals were a 'protestation éclatante' against the feudal system, they signally failed of their purpose, nor could the bishops have been its principal promoters. The feudal system was very far from being extinguished by the rise of Gothic at the end of the twelfth century, and as for the bishops, they themselves were in many cases great feudal lords, not less troublesome to their sovereigns than the lay lords, and equally tenacious of their rights and privileges as against the common people. If, again, French Gothic was 'the exact reflection of the ideas of the nation' early in the thirteenth century, one has to ask, what was the nation? It might have been possible, five hundred years before, to talk of a people, at any rate of a kingdom of the Franks, when Charlemagne ruled from the Atlantic to the Rhine and from the Mediterranean to the North Sea; but in the latter part of the twelfth and early part of the thirteenth centuries there was not one French nation, if such a term is permissible at all, but several. One has to ask also what we are to take as typical French Gothic? If the cathedrals of Rheims and Amiens are the type, those of Angers and Poitiers are not. The various forms of Gothic in France itself are much too diverse to admit of being classified and explained in any such delightfully simple manner. Moreover, there is English Gothic to be taken into account, more national, not less characteristic of its people, equally entitled to be designated Gothic. Even if Viollet-le-Duc's account of it explained correctly how from contemporary causes there came about this great outburst of building, it in no way explains why its architecture should have taken the line that it did, why it should have run out into multiplicity of detail, infinite caprice, and fantastic invention, as against the instinct for solidity, breadth of treatment, rhythm, and proportion that had characterized the architecture of the older world. I am not here advocating one manner of architecture as against another ; all that I am endeavouring to do is to seize certain general characteristics of medieval architecture to show that they were different from those of earlier art (whether better or worse is immaterial to my argument), and to put before you the problem of their origin.

For nothing quite like it had happened before. Hitherto the line of descent of Western architecture had been direct and consecutive ; from the earlier civilization of Egypt and the eastern area of the Mediterranean to Greece, from Greece to Rome, from the Western Empire of Rome to Romanesque with strains from the north and east at each successive stage. But now, in the latter part of the twelfth century appears this new spirit, transmuting the feeble legacy of Roman architecture into an art of extraordinary beauty, giving it a vitality so intense that out of it sprang into splendid life the art of the sculptor, the glass painter, the worker in metals and all the kindred arts, forming out of it an art that did undoubtedly express the inner life of its time. We shall have to go back to times and regions scarcely within the range of history if we are to find a clue to the instinct that lay at the back of this astonishing art, so diametrically opposed, in its spirit, to the classic art of the ancient world, this art that found for itself its own ideal of beauty not only in architecture but also in sculpture, for though we talk vaguely in the same breath of the finest Greek and the finest Gothic sculpture, we are in fact mixing up two totally distinct types, separated by the gulf that lies between the beauty of the north and the beauty of the south. We may find that the ultimate origin of Gothic was not national but racial.

North-western Europe, or, to be more accurate, North-western France and England, is the home of Gothic architecture. Both countries, before and during the Roman occupation, were inhabited by Celts. In both, after the Romans had gone, the Celts remained, more enlightened and far more civilized than the fierce conquerors from the north and centre of Europe who succeeded the Romans. I deal with France as the simpler illustration of my argument. In the first dawn of history we find three races in what is now France ; the Iberians, that mysterious people traces of whose art are still to be found in Spain ; the Ligurians in the Rhone basin ; and lastly the Celts, themselves invaders from the heart of Europe. Besides these races there had been settlements in the south of Phoenician traders, and of the Greek colonists who founded Marseilles. Of these elements the Iberians barely survived in the Basque peoples, the Greek colonists were lost among other races, the Ligurians were conquered by the Celts, and the Celts became the typical and predominant people of Gaul, in touch with their kinsmen in Great Britain, and, as in the case of Brittany, largely reinforced by immigrants from the other side of the Channel. The Celts in their turn were conquered by the Romans, but the Roman conquest was in general humane. The inhabitants settled down with their conquerors, learnt much of their civilization and prospered greatly under it. As the Romans found them, they described the Celts as ' courageous, intelligent, sociable, and demonstrative ', but, on the other hand, vain, fond of display and ornament, uncertain of purpose and destitute of any idea of order. Some of these characteristics, if one may say so without incivility, have been known to appear again in later history. The superb political qualities of the Roman gave this gallant and quick-witted people the stability that it lacked, and Gaul became the best romanized and the most civilized of all the provinces of the Empire. It readily assimilated the political methods of Rome, its habits of life, and its architecture as the expression of those habits, and so long as the Western Empire lasted, the architecture of Gaul, in so far as it had any architecture, was the architecture of the Empire, gradually degenerating in technique even if it might be gaining in freedom.

Meanwhile, immense and obscure movements of peoples from east to west had been occurring periodically in the centre of Europe. Just as the German tribes had pushed the Celts westward in the third century[1] B.C., so, in the fourth century A.D., the coming of the Huns stirred up the wasps' nest of the Goths, and sent them buzzing over Southern and Western Europe. The two

[1] It is supposed that the Celts moved westward into what is now France in the 7th cent. B.C., and to Britain in the 3rd cent. B.C. See *Cambridge Ancient History*, vol. ii, chap. ii. 33–5.

great powers of the Visigoths and Burgundians established themselves in Gaul, leaving unconquered Brittany and the Saxon fringe along the north-west coast, and we have now, in the middle of the fifth century A. D., reached the real period of the Dark Ages, the beginning of that long era of barbarism and bloodshed out of which modern Europe has developed. Aëtius, the last champion of Roman Gaul, died in 454, and the Huns were in Gaul in 451. The Franks succeeded the Burgundians and Visigoths, Clovis forced his way to the front by the murder of his rivals and so founded the kingdom of the Franks and that Merovingian dynasty which, after two hundred years of bloodshed and treachery, made way for the splendid dominion of Charlemagne. I need not follow further the course of general history; the point that I wish to bring before you is that, when the Western Empire fell, the Celtic inhabitants of France were relatively speaking a highly civilized people. After the retirement of the Romans the Celts were in the position of a subject population. France was overrun by a succession of savage Teutonic races, under whom the arts and civilization built up under the Roman occupation gradually disappeared. Scholars and artists fled the country and found refuge in Ireland and elsewhere, and curious evidence of this was found in a sentence embedded in a sixth-century Latin glossary. Referring to the Huns, the writer says that on the coming of the Huns, the Vandals, the Goths, and the Alani, ' omnes sapientes cismarini fugam ceperunt, et in transmarinis, videlicet in Hibernia et quocumque se receperunt, maximum profectum sapientiae incolis illarum regionum adhibuerunt'. Ireland became for a time the refuge of civilization and the arts, and Gaul was left to the tender mercies of the Barbarians. When the Huns came, towns were hastily fortified with anything at hand, the stones of public buildings, temples, monuments, and sarcophagi. The Bishop of Auch, writing about 450, and possibly using the term Visigoths for barbarians in general, says that ' everywhere after the coming of the Visigoths was death, destruction, massacre, and fire'. Euric (died 485), the King of the Visigoths, of set policy let the churches perish. The Arts fell away into the abyss of the Dark Ages, and architecture was not to emerge again till the age of Charlemagne, and then only in timid imitations.

We are now at the beginning of what is generally called Romanesque, that great chapter of architecture which started with what was left of the tradition of later Roman architecture, the round arch and the column. Handled at first by the barbarians in a crude and tentative manner, it gradually advanced both in design and construction till it attained the great quality of such a church as Vézelay. Its ornament was barbaric, consisting of chevrons, interlacements, and the like;—symbols of ornament rather than ornament itself—but these men had definite architectural instincts, they cared little for detail, what they aimed at was a massive strength that would withstand not only the fire and sword of the invader, but the ravages of time. The energy, the ferocity, and the directness of Clovis, of Charles Martel, of Charlemagne, of Foulque Nerra of Anjou, of Eude of Blois, of

Robert le Diable of Normandy seem to find their true expression in these tremendous buildings destitute of any grace, without any subtlety of fancy, but stark, strong, and enduring. Instances are to be found in nearly every part of France ; the gloomy Abbaye aux Hommes at Caen, the strange church of Tournus in Burgundy, Vezelay, and the huge establishment of Cluny, now destroyed, the churches of the Auvergne, Issoire with its apses and its towers, Le Puy with its vast flight of steps and its inlays of lava ; the church of St. Sernin at Toulouse, of Montmajour, St. Trophimus at Arles, and St. Gilles in Provence, all these in their different ways are typical of the spirit of Romanesque architecture. Grim and ferocious even to barbarism in the motive of their design, barely hinting in such churches as Vezelay at the imminent change, they show the same elements, the rudimentary vaults, narrow openings, piers and walls of disproportionate strength, the strange barbaric ornament, an art, even when touched by Eastern influence, bound by conventions ill understood. Here and there the Byzantine influence appears predominant, though in the native idiom, as in the church of St. Front at Périgueux, and St. Pierre at Angoulême, but these are isolated examples in regard to the rest of France, for the extraordinary thing is, not that these examples should be found at all, but that in spite of the intercourse between East and West there should have been so few of them, and that Western architecture, apart from sculpture, was so little affected by the art of the Byzantine Empire. It would almost seem that the rude successors of the Celts were incapable of understanding the subtle art of Byzantium.

Within a short half-century all this was changed. Within fifty years of the completion of the churches of Périgueux, of Angoulême, and Vezelay, we find an entire revolution in architecture, reaching that perfect expression which is to be found at Rheims, or at Bourges, buildings which imply not so much a development on existing lines as a *volte-face* in artistic outlook—the search for height, and the effort at expansion outwards and upwards instead of the low-browed gloom of the Romanesque church, the passionate quest of freedom and individualism, the caprice and poetry, the mysticism, it may be, which had been thrust into the background, not only by wave after wave of barbaric invaders, but also before that era by the iron heel of Rome. I suggest to you that at the date of the rise of Gothic architecture in the twelfth century there was one racial element only, among the many that went to the making up of the French people, to which the movement responded, and of which it might be taken to be the expression. It was not inspired by the Roman, who had aimed at other things, nor by the Visigoth, the Burgundian, the Frank, the Saxon, or the Norman, fighters, adventurers, and men of action. The Celt alone remains as the strange spirit in whose dim instincts might be found the source of the new art, its poetry, its mysticism, its caprice. I am not, of course, talking of the technical development of Gothic architecture: that can be explained; the steps by which each advance was made can be easily traced, and in this sense there is no break between Romanesque and Gothic. But what has to be explained, and must always remain to a great extent a mystery, is the change in the orientation of architecture within the short space of, say, three generations ; a change so radical that for the next three or four hundred years artists were to make for qualities in architecture and the arts, not merely dissimilar but even to some extent antagonistic to the qualities of all phases of art hitherto known in the Western world. Political causes contributed. The persistent efforts at ecclesiastical and religious reform, the growth of the monarchy from the position of simple predominance among the feudal lordships into something like real sovereignty, the emancipation of the lower classes, the efforts made by the *bourgeoisie* to establish their rights individually and collectively, the rise of the lay element, the intellectual and literary renaissance of the twelfth century, all these are of course of great historical importance, in so far as they represent the movement of ideas surging in the twelfth century, and as a friend [1] has suggested to me, the coming of age of the middle class which must have been mainly Celtic in origin. But they are not so much causes of the change as indications of the renewed intellectual activity, of which the change itself was an expression. One may admit that these movements gave the occasion, but they do not explain the transformation of the inner spirit of architecture which I have attempted to indicate, and I suggest to you that that inner spirit is the expression of the true autochthones of France,

[1] The late Sir George Prothero.

of that Celtic race which had fused itself with
Rome in the happy days of Gallo-Roman France,
which had been trampled under foot by Visigoths
and Franks, and the pirates and freebooters of the
north, but which had never wholly died out, and
which had in it sufficient vitality to find itself again
under the Capet dynasty. One must recollect
that Romanesque was not the architecture of the
Celt, who during these stormy times was in the
position of a conquered people. It was the architecture of the half-civilized conquering races that
overran France, all of them Teutonic in origin,
each attacking in turn this problem of architecture
with the naïve instincts of the barbarian, and doing
the best it could with the legacy of Rome. The
quality of Romanesque, the appeal that it makes
to us in its massiveness, its attempt at symmetry
and rhythm, in its rudimentary methods of proportion, has to be judged from the standpoint
of Roman Classic, with due allowance for the
strong infusion of the half-savage instincts of the
northern races. It is impossible to apply that
standpoint to Gothic architecture. We are here
in the presence of a new spirit, expressing itself
in a manner which reverses the ideals, and turns
its back on the procedure of classical art. We
cannot look to Greece or Rome for its inspiration.
We cannot find it among the barbarians. There
remains the Celtic race, long in abeyance, never
wholly extinguished, still vital to-day. The
qualities of that race are very unequal; but
history shows that it has possessed high artistic
instincts, in certain directions, and a capacity for
spiritual exaltation, such as might fitly express

itself in that austere art, in that chastity of thought which stamps Gothic architecture of the twelfth and thirteenth centuries with ineffable distinction. The rise of what we call Gothic in the latter part of the twelfth century is not to be explained as a protest against feudalism, its meaning is not exhausted by expositions of its technical development as an affair of vaulting and fenestration. I suggest to you that, roughly speaking, it represents the re-emergence of the Celtic temperament, the articulate expression of the instincts and ideals of one of the oldest and most vital elements in the long pedigree of France. It is in this way that the rise of Gothic architecture is a tremendous example of atavism, a throwback, or to put it another way, the re-emergence of long-lost instincts and tendencies. These instincts will gradually lose their power, before a new and compelling force, the Renaissance. The Celt, by then, will have had his day, and the ideals of an older civilization will recover their ascendancy. Two hundred years later the pendulum will swing again, and we shall have the Romantic movement, already drawing to its close. It is in this action and reaction of deep-seated forces, these constant groupings and regroupings of the elements of race, that the key should be sought to the changing cycles of ideas in modern art and literature.

That this theory of Gothic architecture, this attempt to get to the back of a great historical development is no more than hypothesis, I admit at once. When we come to the Renaissance we are on surer ground. Here there was a deliberate return to the methods of a forgotten civilization.

It was a movement started by scholars, but so quickly taken up by artists and laymen that within a hundred years the art, not only of Italy but of Western Europe, was revolutionized again. I need not pursue this familiar aspect of the Renaissance. The point to which I would draw your attention is, that where imperial Rome had been strongest the return to its art in the fifteenth, sixteenth, and seventeenth centuries was most complete. In Italy it came naturally, and almost as a matter of course. In spite of Huns and Vandals, Lombards and Franks, the Italian never wholly lost touch with Rome, or forgot that in some obscure way he was the inheritor of her greatness. The Gothic of the north was never at home in Italy. It only needed the scholarship of the Humanists, the work and example of such men as Alberti, to rouse long dormant instincts into brilliant life, and to start again that Roman architecture which has dominated modern Europe. Of countries outside Italy France was easily first in the field. The effort of Henry VIII to introduce the Renaissance directly into England was diverted into the wrong channel by political causes which set it back for 100 years, whereas, in France, the enterprise of François I started a movement which lasted without interruption till the end of the eighteenth century. In a less complete degree than France, and with wide variations, the art of the Renaissance established itself in Spain. In Germany, always less dominated by Rome than France or Spain, the Renaissance appeared in a mutilated form, set forth in text-books of the orders, adorned with all the exuberance of Teutonic fancy. That the Renaissance should have won its way in Italy at once, and without let or hindrance, is intelligible ; that it should have done so in France with the rapidity and permanent results that it did, points again to some predisposing quality in her people, some latent instinct for order, symmetry, and restraint, which was roused to reassertion by the extravagance of later Gothic. Gradually the art of France freed itself. Goujon and his contemporaries cleared the air of the morbid imaginings of the tombmakers, architecture more slowly shook itself free of the tangle of details in which the art of the end of the fifteenth century was losing itself, and finally found that true Roman manner of design which had been the goal of all students since the beginning of the sixteenth century.

Again I suggest that out of that welter of peoples, Celts, Romans, Goths, Burgundians, Franks, and Northmen, some instinct from an earlier age had emerged, that just as Gothic may have risen from the element of the Celt, so some dim instinct left by Roman civilization and by the mixture of the Roman with the native population, may have opened the way for this vast change in the direction of architecture, a change that went back on all that had been done in France for some 500 years.

In the history of the Arts, particularly in architecture and sculpture, instances of atavism constantly occur in detail. I will only call your attention to one rather sinister example, and that is the feeling for the ' Macabre ', for the horrible for the sake of the horrible, which appears from time to time in Western art. The 'Macabre', in the sense in which I am using the term, has to be differentiated from the grotesque, from hieratic representation, and from simple realism. The grotesque I understand to be, in the main, the presentation of some humorous aspect of things which has caught the artist's fancy, and which has impressed him so strongly that he reproduces that aspect for the sake of its humour, accentuating and exaggerating it in order to keep a firm grip of its essential character, and to drive his impressions of it home to the hilt. The grotesque does not aim high, but it has its place in art, as part of the general version of humanity. Or again the artist may use a version of the macabre for an ethical purpose. There was a notable instance of this in the recent work of M. Raemaekers. The superb drawing and movement in the cartoon of Germany's dance with Death justify it on technical grounds, and its high moral purpose redeems it from the distinctive taint of the macabre. Hieratic representation in art again may be horrible and repulsive, but repulsiveness is not the motive of the artist's work. He presents it because he has to, as a symbol of religion, such as the monster figures of Indian gods, or as part of a story such as the Medusa in the metope of Selinus, or the Harpy on the tomb of Xanthus ; or as a recognized convention, such as the masks of the Greek drama, or certain figures on Greek vases. Simple realism explains itself. In the wild bull crumpling up his pursuers in the Vaphio vase the artist has faithfully represented an incident that he had seen himself. All and each of these different phases of art proceed from a different motive from that of the macabre, which craves after the horrible for its own sake, and very rapidly becomes one of the worst degradations in the whole range of art. Every one who has seen the Romanesque churches of Italy and Southern France will recollect the monsters that stand by their entrances, lions devouring men and animals, such as those on the side of the entrance to St. Trophimus at Arles, or the terrific figures along the west front of the abbey church of St. Gilles. Where did this strange violent motive come from? Signor Rivoira says it was derived from Syria and Chaldea;[1] but it is found in almost identical terms in an ivory of the seventh century B.C., of a lioness devouring a calf from the temple of Artemis Orthia at Sparta. It is found again in the Corbridge lion. Are we to take all these instances as versions of an Eastern motive? Should we not rather look upon them as the irresistible expression of a common instinct that had its origin far back in the home of the northern races, in the savage gloom of some Hercynian forest, an instinct that struggled for utterance in the Dorian in Greece, in the Lombard in Italy, in the Visigoth in Provence, in the sculptor on the far Northumbrian border? Professor Haverfield has attributed a Celtic origin to the Corbridge

[1] In the British Museum there are two stone lions with Hittite inscriptions which bear some slight resemblance to the Lombardic monsters, and these may lend some colour to Rivoira's view. My point is that, whatever its origin, this motive vanishes in Greek and Roman civilization, but reappears with the advent of the invaders from the north.

lion ; but surely its affinity is with those far away instances that I have noted. These artists could not have copied each other, nor is the motive of the monster and victim on the same footing as mechanical details of ornament, such as the egg and dart, the fret, the wave, and the like. It is more serious, more deep-seated, the expression of an instinct and temperament which is racial and not local. The same instinct for the horrible and cruel appears in some of the Last Judgements in the Tympana of medieval doorways. In many of them the demons are merely humorous creatures, who cheerfully shepherd the souls of the damned into the pit. But here and there the true macabre appears. In the Last Judgement on the west front of the cathedral of Autun, a serene and mighty Angel holds the scale, but pulling at one of the bowls are two nightmare figures, bony, horrible, misshapen, immensely tall, writhing in unutterable contortions. Such figures are rare in earlier work, but as Gothic art failed of its high purpose this morbid tendency forced its way to the front. French sculptors of the twelfth and thirteenth centuries had produced figures of splendid dignity, the fit expression of all that was noblest in their generation ; but in the latter days of Gothic, just as the high ambition of architecture gave way to technical vanity, so the art of the French and Flemish Primitives reveals their morbid pleasure in the agonizing details of martyrdom ; and whereas the sculptor of the thirteenth century could create that splendid lissom figure of Samson in the west door of the cathedral of Auxerre, the sculptors of the end of the fifteenth century were falling into the art of the charnel house and the grave. The Fleming and Burgundian drew their inspiration from Germany, not from Italy, and it was thus that this most vicious instinct came again to the surface, an instinct which elsewhere and before the war broke out I had called ' Teutonism ' in art.

The entombments of Solesmes, of Tonnerre, and Chavoure, the naked figures of corpses on the tombs of François I and Henri II at St. Denis, culminate in the horrible figure of Death on the tomb of Renée of Chalons. The medieval sculptor at Autun, Ligier Richier with far less excuse at Bar-le-Duc, used their art to call up visions of all that is most vile and terrifying to mankind, and the danger of the macabre in art lies in this false appeal, this effort to arrest attention by means that finer spirits repudiate. It is an instinct that constantly asserts itself. It is found in the art of Italy and Spain in the seventeenth and eighteenth centuries. It is to be found to this day in a certain type of picture that appears year after year in the Salons. It inspired the odious presentation of a great Greek tragedy a year or so ago. It may possibly explain the blatant developments of modern German music.[1] Closely allied to it is a recent movement which affects to ignore all the aesthetic sensibilities built up since the earliest days of the art of the Western world, offers us instead pseudo-scientific researches into the peculiarities of vision of the artist himself, and seeks to arrest attention, not only by the choice of repulsive subjects, but also by presenting

[1] Written before the war.

them in the manner best calculated to hit the spectator by shocking both his moral and his common sense. Beneath this false art lies that craving for sensationalism which has prompted the worst excesses of the macabre in art. Throughout the history of art, as there have been men who sought only beauty, so there have been those who either revelled in horrors or took a short cut to notoriety by their representation, and it is possible that this morbid tendency is the survival or half-conscious reminiscence of the ferocious instincts of a violent race. On the other hand, the unrecognized survival of deep-seated instincts may partially explain the rise of the artist of genius, out of space as it were, those cases where there seems nothing in his heredity or surroundings to account for his existence. At least one would rather find the explanation here than in merely physiological accidents.

I admit that these suggestions are only speculation, and to work them out in detail would involve wider knowledge than I have any pretence to claim, and greater leisure than is possible to a professional man ; but I offer them for what they are worth, because I am convinced that the whole explanation of any great movement in Art, such as the rise of Gothic on the one hand, and of Neo-Classic on the other, is not to be found in merely technical developments. That materials and climate, social, political, and local conditions, manners and customs have had a profound influence in the shaping of architecture one admits at once. It is comparatively easy to trace the technical development of Western architecture. Yet however ably this is done the final mystery remains where it was. These explanations will not tell us what was at the bottom of it all, what spirit inspired and expressed itself in Gothic on the one hand, Neo-Classic on the other. If we are really to understand the meaning of great movements in thought we must make the attempt to drive back to these root instincts of our race. In the last of his writings a famous French surgeon (Dr. Lucas Champronnière) said that our thought and our discoveries are frequently merely ' the *résumé* of observations of the past, not only of that past from which we directly derive our instruction, but of *a past of which we have no knowledge* '. It is in this past of which we have no conscious knowledge, in instincts long forgotten yet dormant only, that the clue should be sought to the mystery of the great historic movements of art.

THE BRIDGES OF LONDON [1]
1815–1920

BRIDGES over running water seem to possess a certain vitality peculiar to themselves. Whether this impression is due to the springing arch, or to the suggestion of being above the earth and to that extent aloof from it, or to the movement of the water, they affect one differently from other buildings, and from time immemorial seem to have had an irresistible attraction for mankind which it is not easy to explain. Why, for example, in medieval times, should people have insisted on building their houses on bridges, in spite of the extreme inconvenience to themselves and others, the risk of fire from their ramshackle houses, such as the disastrous fire on old London Bridge, when three thousand people were said to have perished, the dangers to the fabric of the bridge and other common-sense reasons, all of which were impartially ignored, in some vague anxiety to put as great a distance as possible between one's habitation and the noisome things of earth ? Why, again, was the head of the Roman College of Priests called the 'Pontifex Maximus' ? Was he a bridge-builder to heaven, like the Pontifex Maximus of later date, or was he in the first instance a builder of actual bridges, the man of skill and genius who met and conquered the forces of nature for his fellows ? We must leave these questions to the

[1] (Issued in *London of the Future*, by the London Society: Fisher Unwin.)

student of primitive religion and content ourselves with the fact of the indefinable fascination of bridges over running water, and one other fact, that in the twelfth and thirteenth centuries the ' Frères Pontifes ', the brethren of the order of S. Bénézet, were actually the men who built and maintained the few bridges that existed. The great bridge at Avignon, 900 metres long, now in ruins, the Pont Saint-Esprit across the Rhône, 919 metres long,[1] still in use, though much altered, were carried out by these indomitable brethren. Probably old London Bridge, built in the last quarter of the twelfth century and attributed to Peter the priest of Colechurch, was the work of the Brotherhood. From the clergy, bridge-building passed into the hands of the military engineers, but its design was recaptured by architects at the Renaissance, and remained with them till the middle of the eighteenth century, when it passed into the hands of the specialized engineer. All the finest stone bridges of the seventeenth and first half of the eighteenth century seem to have been designed by architects, and it is a matter for regret that the designing of bridges should ever have passed out of their hands, but architects were themselves partly to blame. That able impostor, J. H. Mansart, undertook bridges, as he was ready to undertake anything else ; but the total collapse of one of his bridges within a very few years of

[1] The length of Waterloo Bridge between the abutments is 1,240 feet ; Westminster Bridge is 810 feet long between the abutments, little more than one-fourth of the length of the Pont Saint-Esprit. The Pont Saint-Esprit is about 40 kilometres above Avignon.

its being built was one of the reasons that led to the establishment of the ' Ponts et Chaussées ' in France and its complete reorganization by Perronet in the middle of the eighteenth century.[1] To Perronet and his school were due some of the finest bridges in France; but the specialization of construction, the severance of engineering from architecture, had already begun, with disastrous results to both, and the process was completed by the introduction and development of iron, and later of steel construction in the last century. One could wish for a class of Frères Pontifes amongst us now, men equally conversant with construction and design, men capable of getting the utmost possible out of material, both for scientific and aesthetic purposes, who out of their construction would evolve forms that are beautiful to look upon, instead of plastering ornament on to their construction. For a bridge-builder it is not enough to be a master of construction and building processes. 'These ought ye to have done and not to leave the other undone ', and that ' other ' is of the very essence of the work—the grasp of the imaginative problem as a whole, the realization of what the bridge means, not merely as a means of transit but as a symbol of the life and civilization of the people who use that bridge ; and it is here that our modern bridge-builders fail so lamentably. Some years ago it was necessary to construct a bridge above the Pool of

[1] The last and perhaps the finest of architects' bridges is the Ponte S. Trinita in Florence as rebuilt in 1769. The bridge has three half-elliptical arches. The centre arch has a span of about 116 feet, and the piers are about 18 feet in width transversely.

London, the very centre of business of the waterway of the greatest city of the world, and all we could produce was that monument of artistic ineptitude, Tower Bridge, with its towers like a confectioner's cake and the clumsy curves of its suspension bars. I make no criticism on the engineering solution of the problem. It answers its purpose efficiently, and is no doubt very well done, but the aesthetic result is patent to any one who looks eastward from London Bridge. What is forgotten nowadays in dealing with monumental problems of design, such as bridges and the like, is that, after all, technical knowledge of construction, though indispensable, is not the whole of the equipment necessary. Imagination, passion, fine ideals, and a range of thought that lives among the higher spaces of life are equally vital, are in fact the essential basis of great design. Our bridges, serviceable enough as a means of transit, have a distressing habit of lapsing into bathos. The artist is wanted here not less than the scientific constructor. Where the latter sees only his calculations and formulae, the artist will see possibilities of emotional expression. He is trained in the appreciation of form, line and mass, in selection, in sacrifice ; and it is his business to interpret the aesthetic qualities that lie latent everywhere, not by superadding things that have no relation to the essential purpose of his subject, but by searching out the beauty that is inherent in it. The best solution, no doubt, would be to combine the two faculties in one man ; but if that is not possible in view of the intricate complexities of modern construction, the engineer and the artist might at least co-operate. After all, a bridge is about the most prominent object it is possible to construct anywhere, and as it is impossible to escape it, the aesthetic effect of the bridge must be a vital part of the problems of its design. There ought to be no difficulty in the co-operation of engineers and architects if both of them know their business and if only they are ready to pull together.

It is not very easy to say where the London bridges begin. They end with Tower Bridge, but up the river the interminable houses extend on both sides up to Putney, with the welcome break of Hurlingham. Strictly speaking, the London bridges extend from Tower Bridge up to Hammersmith, and of these, Tower Bridge, London Bridge, Southwark, and Blackfriars are under the control of the Corporation of the City of London, and Waterloo Bridge, and the bridges westward, up to and including Hammersmith Bridge, are under that of the London County Council. With the exception of Westminster and Chelsea Bridges, which were built for the Government, all the L.C.C. bridges were in the first instance built and maintained by private companies under powers conferred by Acts of Parliament, and the cost was to be recovered from the tolls. In 1877 an Act was passed empowering the Metropolitan Board of Works to buy out these interests and free the bridges to the public. This was done at a cost of £1,376,825, the control passing finally to the L.C.C. in 1895.[1] It must be admitted that

[1] See the *Report on Bridges*, printed for the L.C.C., 1914, P. S. King & Son, 2 and 4 Great Smith Street, S.W., to which very useful pamphlet I am greatly indebted.

THE BRIDGES OF LONDON

with all its faults of omission and commission the old Board of Works rendered two admirable services to the public—the freeing of the bridges and the construction of the Victoria Embankment. Unfortunately, the bridges so acquired displayed a want of taste and knowledge inconceivable in any period but that of the second half of the nineteenth century. To those who designed them these bridges seem to have had no symbolism, they were just a means of getting from one side of the river to the other; had they been content to leave it at that, one would have gladly acquiesced in their bare construction, but the municipal authorities of the time seem to have thought it necessary to make some concession to aesthetic demands, and instead of consulting some competent artist, the engineer, with sublime self-confidence, launched out into uncharted seas, and produced the abominations of the Hammersmith Suspension Bridge and Chelsea Bridge. The L.C.C. *Report on Bridges* says that Hammersmith Bridge was the first suspension bridge over the Thames. It was designed by Mr. William Tierney Clarke, and was opened on October 7, 1827. When the Metropolitan Board of Works acquired it in 1880 the bridge was found to be unsafe and too narrow. A new bridge, opened in 1887, was designed by Bazalgette, and the *Report* continues : ' The only portions of the original structure which were allowed to remain were parts of the towers below the road and the abutments.' The old masonry towers were replaced by ' lighter ones of wrot iron '. I do not know what the old ' towers ' were like ; they are described as having had openings for traffic only 14 feet wide, and were no doubt extremely inconvenient, but anything more mean and commonplace than Bazalgette's 'lighter ones of wrot iron' it would be difficult to conceive. They are as bad as the deplorable towers of Chelsea Bridge, two of the worst eyesores in the whole length of the river. If only the engineers had learnt to leave well alone!

Battersea Bridge seems to be a reasonable limit for the London bridges. To the west of it the river sweeps round Chelsea Bay and then turns southward in a long reach to Wandsworth Bridge. Moreover, older London stops at Chelsea and Battersea, with their memories of the eighteenth century, and the old bridge of Battersea will always live in the work of a great modern painter. The old bridge was constructed for Lord Spencer in 1771–2. It was formed entirely of wood in nineteen spans, and, after having valiantly done its work for a hundred years, it was purchased by the Metropolitan Board of Works, who found its condition so unsafe that they had to set about rebuilding it at once. Bazalgette designed the new bridge, which was opened in 1890. It is a singularly ugly structure. Five segmental arches of cast-iron ribs on stone piers span the river, and above the arches is a large cove which the engineers believed would give 'lightness to the design of the bridge'. Its result is to make it look weak, and the L.C.C. *Report* says that on foggy nights it has led to accidents, because bargemen mistake it for the outline of the arch. Above this cove there is a preposterous little balustrade

THE BRIDGES OF LONDON 101

of a Moorish design, reminiscent of the Alhambra —the Alhambra, that is, of Leicester Square, not of Granada. Altogether it is a poor design. The work of the engineer of the Metropolitan Board of Works was, to say the least of it, puzzling. Some of it was very good, as, for example, the Victoria Embankment,[1] and some of it very bad, as, for example, the Battersea Bridge and the Chelsea Embankment.[2] The difference is so great that it seems inconceivable that they were all designed by the same hand, and that the man who could design the splendid detail of the Victoria Embankment could also have been responsible for the coarse and ignorant detail of Battersea Bridge, or, for the matter of that, of the Chelsea Embankment. The details of the latter are very inferior. The mouldings are ignorant and have no meaning, and the engineer perversely rusticated the retaining walls of the embankment on the side of the river, thus affording a convenient resting-place for all the garbage of the river. Battersea Bridge has, however, one merit. It is approached by long straight roads, which, though not quite axial on the north side, are nearly so, and, as the gradient of the bridge is low, a fine long vista is obtained across it southwards.

There is little to detain us in the Albert Suspension Bridge, hung like a great spider's web spun across the river. It was designed by Mr. R. W. Ordish, C.E., in 1873, and the only remarkable thing about it is the width of the

[1] Opened in 1870; J. W. Bazalgette, engineer.
[2] Opened in 1874; J. W. Bazalgette, engineer.

centre spans of 383 feet and the height of the towers, 101 feet above high water. The *Report* says, ' The whole structure is from an engineering point of view very unsatisfactory ', and from an artistic point of view it is worse than that. It looks like a temporary gangway flung across the river, and in fact is not very much more. Both on this and the Chelsea Bridge, troops have to break step in crossing, and no load above five tons is allowed.

The Chelsea Suspension Bridge is even worse, because it is more ambitious. This bridge was built in 1851-7 from the designs of Mr. T. Page, C.E. Its kiosks and gilt finials, its travesty of Gothic architecture in cast iron, its bad construction and its text of ' Gloria in excelsis ' above the arch between the piers are redolent of 1851, the year of the Great Exhibition, the *locus classicus* of bad art and misguided enthusiasm. In spite of its pious aspirations, the bridge had to be strengthened six years after its completion and again in 1880, and even so will not carry more than a load of five tons. The worst of it is that this bridge blocks the view of Chelsea Hospital as one enters London by trains crossing Grosvenor Bridge. Instead of wasting money on paint and patching, it is to be hoped that this bridge will some day be replaced by a permanent structure. The approach on the south side next Battersea Park is a fine one, and that on the north side could be improved without much difficulty. This matter of the approaches to our bridges has been far too much ignored from the very first. It is a point that was never overlooked by the best French bridge-builders.

The next traffic bridge down the river is the important bridge of Vauxhall. This is so far [1] the last completed of the London bridges, and though it is open to criticism, it is notable as being the first serious attempt in recent times to regard a great bridge as something more than a mere engineering problem. There was an earlier bridge at Vauxhall, built in 1811-16, at a cost of £296,998. This bridge, designed by John Rennie, consisted of nine iron arches of a span of 78 feet, and was the first iron bridge across the Thames. It was taken over by the Metropolitan Board of Works at a cost of £255,000, but a few years later was found to be unsafe, and in 1895 the L.C.C. obtained powers to rebuild the bridge, and the work was completed in 1906 from the designs of Sir Maurice Fitzmaurice and Mr. W. E. Riley, superintending architect to the L.C.C. The bridge is in five spans, with steel arches on granite piers and abutments. A large ovolo moulding projects above the crown of the arches, carrying a light iron balustrade. On the piers above the cutwaters are steel recessed panels, four on each side, filled with bronze figures of heroic size by Mr. Alfred Drury, R.A., and Mr. F. W. Pomeroy, R.A., representing Local Government, Education, Science, Fine Arts, Pottery, Engineering, Agriculture, and Architecture. The bridge is of fine width—80 feet. In order to get headway without too steep a gradient, the arches had to be kept very shallow, and it was to correct this thinness that Mr. Riley designed the unusual balustrade iron bars with a secondary rail above.

[1] 1920.

Though perhaps the treatment is rather light, and hardly in scale, it is a metal treatment, and a break away from the habit of trying to get the effect of stone in cast iron. The figures themselves are dignified and impressive, but their position on the piers is open to question. It is doubtful whether in any case it is right to put figures below the bridge level and on the face of the piers instead of on the top of them. In this position it is difficult to get them into scale with the bridge itself, as can be seen from the failure of the figures on the Pont de l'Alma at Paris, where the effect is almost grotesque. In the Vauxhall Bridge the setting is hardly adequate for the figures, and the only people who can see these figures are the crews of the tugs and barges that go up and down the river, as the Embankment stops short on either side, and the nearest point from which the bridge can be seen is the Embankment in front of the Tate Gallery. Moreover, the figures, being of bronze against a dingy green paint, are lost against the background. When the sun is west of the bridge, these figures are hardly visible at a distance, and it ought to be almost a rule in our climate that when bronze is used in the open it should be set against a light background, such as Portland stone. Here, however, at any rate, a serious effort has been made to treat the bridge as a great public monument.

There is nothing to detain us on our way downstream till we come to Westminster, for the Lambeth Suspension Bridge is an insignificant affair; its end bays are sagging seriously, and for over twenty-five years it has been practically condemned. In an important part of the river, such as this, a suspension bridge should never have been allowed. Looking down the river from Vauxhall Bridge the outline of the Lambeth Bridge cuts the lines of Lambeth, St. Thomas's Hospital, the Houses of Parliament, and Somerset House beyond in the most disagreeable manner ; but the bridge-designers of the last century seem to have thought it unnecessary to consider the bearing of their design on its setting and surroundings.

With Westminster Bridge begins that splendid series of bridges, embankments, and buildings which makes the view from Westminster Bridge the finest thing in London, perhaps in the world. The old Westminster Bridge was begun in 1739 and completed in 1750, from the designs of Labelye, a Swiss engineer, at a cost of £389,500. Labelye seems to have been careless or over-confident about his foundations, for he omitted any piling under the piers and built them in caissons, directly in the soil, at a depth of from 5 feet to 14 feet below the bed of the river. The piers were constructed of huge blocks of Portland stone, weighing from one to three tons and fastened with iron cramps. There were thirteen large arches, the largest 76 feet in span, and two small ones at the ends, the total length of the bridge being 1,223 feet and the width 44 feet. It must have been a fine-looking bridge. I saw recently in a print-shop a beautiful aquatint in blue and silver of old Westminster Bridge. The view was taken from the site of the new L.C.C. buildings before the Houses of Parliament were burnt and rebuilt, and shows Westminster Hall rising above the arches of the bridge, and beyond it the towers and roofs of the Abbey. Fine as Barry and Pugin's design is in some ways, it blocks the Hall and the Abbey, and it seems to have lost us one of the most beautiful architectural compositions that ever existed in London. But Labelye's construction was faulty, and in 1831, after the removal of old London Bridge, the increased scour gradually washed away the foundations of Westminster Bridge, until its rebuilding became inevitable. The work was entrusted to Mr. Page, the engineer of the Chelsea Suspension Bridge, and was completed in 1862. It is a great improvement on the Chelsea Bridge, and though at a far lower artistic level is, next to Waterloo Bridge, the most satisfactory of the later London bridges. The bridge is in seven spans, formed with iron arches varying from 94 feet 9 inches at the sides to 120 feet wide in the centre arch. The piers and abutments are of granite, and though the details are rather absurd it is a solid and by no means undignified structure, being of ample width (84 feet 2 inches between the parapets) and having the great advantage of important buildings on either side of the approaches at each end. It is true there is no sort of balance in these buildings ; at the south end, St. Thomas's Hospital and the new L.C.C. buildings are too close in to the bridge, and at the north end the business premises on the south side are unequal to the task of standing up to the Parliament buildings across the road ; but seen from Charing Cross Bridge, with Shaw's splendid

Scotland Yard in the foreground, the bridge and the Westminster buildings make a noble group. The lattice girders of the South Eastern Railway bridge block the view, and, till this is removed, Londoners will not realize what they possess.

Waterloo Bridge is the last of the L.C.C. bridges eastward, and by universal consent is architecturally the finest bridge in England, if not in the world. Its history is curious and characteristic of our English methods. In France a bridge of this importance would, as a matter of course, have been undertaken by the State. In England it was the work throughout of a private company. In 1809 an Act of Parliament authorized the formation of the ' Strand Bridge Company ', with a capital of £500,000, increased in 1813 to £700,000, and again increased in 1816. The bridge was designed by John Rennie ; the first stone was laid on October 11, 1811, and the bridge was opened by the Prince Regent on June 18, 1817, the second anniversary of the Battle of Waterloo. The name had already (1816) been changed from Strand to Waterloo Bridge, and the words of the Act of 1816 are memorable: ' The said bridge when completed will be a work of great stability and magnificence ; and such works are adapted to transmit to posterity the remembrance of great and glorious achievements.' It was therefore decided that ' a name should be given to the said bridge which shall be a lasting record of the brilliant and decisive victory achieved by His Majesty's Forces ', and no monument could more fully express the grim and enduring courage of the British soldier of 1815. The total cost of the bridge and approaches was £937,392. In 1877 it was acquired by the Metropolitan Board of Works for £474,200, when the toll-gates were removed. With the exception of certain works necessitated by the scour of the river in 1882, the removal and subsequent return of the original iron lamp-standards, and the skilfully executed alterations for the tramways on the west side, this great bridge has stood its hundred years without any alteration or failure,[1] and, unlike the new bridges across the Thames, there is no limit to the weight of vehicles using the bridge.[2]

The bridge is so familiar that no description is necessary. It is a standing example of what may be done by a good man with the simplest possible means. There is no ornament, except the modillion cornice and the coupled columns above the cut-waters, but the whole design is so admirably balanced, the proportions are so perfect, the details, simple as they are, so exactly right and so instinct with knowledge in reserve, that criticism is ungrateful to one who day after day has passed under its arches and watched it, immutable, yet

---

[1] Written in 1920.
[2] L.C.C. *Bridges, Historical and Descriptive Notes*, p. 56. The bridge is constructed of granite in 9 elliptical spans, each of 120 feet with a rise of 35 feet; the total length is 1,240 feet, and the width between parapets 42 feet 6 inches. The bridge of Neuilly, constructed from the designs of Perronet in 1772, though in five arches only, has very similar proportions. The arches are 120 feet wide with a rise of 30 feet. The piers are 13 feet thick as against the 20 feet of Waterloo, and the width is 45 feet out to out. It is a testimony to the soundness of the work of the English engineer that, whereas the arches of Perronet's bridge settled some 8 inches after the centres were struck, the arches of Waterloo Bridge only settled 1½ inches.

never the same under the changes of our restless skies, gathering up into itself all the elements of romance—the storm, the sunshine, the power, and the tragedy of London's glorious river. Yet I confess to an incessant curiosity as to who really designed the form and fashion of this bridge. That Rennie was the engineer, and an extremely able one, we all know ; but Rennie also designed London Bridge a few years later, and this is so inferior and his treatment of his iron bridges was so unattractive, that, as in the case of the Victoria Embankment, one cannot help asking who helped the engineer. There is a legend in the Temple that Rennie got the designs from some broken-down architect in prison, and the hand of an architect, and of a very good one, is written all over it.[1] When Waterloo Bridge was built architects were still enthusiastic for the severest forms of Greek architecture. Were any of the leading architects of the time consulted ? Did Soane or either of the Inwoods or Decimus Burton lend a hand ; or do we owe the Sicilian Doric of the columns, the fine and even learned profiling of the cornice, the admirable spacing of the rustications, to some unknown draughtsman, some forgotten and unrecognized genius in Rennie's office ? The motives of these columns in this position and the

[1] It has been suggested that Cockerell, who married Rennie's daughter in 1828, might have designed the masonry, but he must have been a very young man when the designs were being prepared. Cockerell, who was born in 1788, went to Greece in 1810, sailed in a small hired Greek ship from Athens to Lycia in 1811, and was absent in Asia Minor, Greece, Sicily, and Italy till 1817. He might have sent a sketch from Greece, but it is unlikely.

modillion cornice were undoubtedly due to an architect, for in the old Blackfriars Bridge, begun in 1760, Robert Mylne, the architect, had also placed pairs of columns above the cutwaters and a modillion cornice with the architrave omitted above the arch, exactly as in Waterloo Bridge. Mylne had used rather attenuated Ionic columns, and the general scale of the Blackfriars Bridge was inferior to that of Waterloo ; but whoever it was, whether Rennie or his draughtsman or another, the designer took this motive and handled it with the audacity of a master. In one point only the design seems to me open to criticism : the voussoirs in the crown of the arch are not quite deep enough. With one more effort of audacity the designer might have broken through his frieze and carried these voussoirs through to the soffit of the cornice. The central voussoirs appear to be about 4 feet to 5 feet deep. According to Belidor's rules for the depths of voussoirs in an arch of 120 feet span they ought to be 8 feet deep.

With Blackfriars begins the series of City bridges. The existing bridge is a standing example of foolish ornament. About forty years ago these stumpy little columns, about two to three diameters in height, with their enormous capitals doing nothing, were rather the fashion, apparently an attempt to catch the Romanesque manner of Burgess. Here they carry nothing, nor have they any relation as a motive of design to the iron arches. In the old Southwark Bridge, now destroyed, with its three great iron arches and massive stone piers, Rennie managed the combination perfectly well, but modern engineers will never learn to let well alone. Tower Bridge is an even worse example than Blackfriars of the same failure in ideas, though candour compels me to admit that the Gothic towers and gateways were designed by an architect. Since the building of St. Paul's no finer chance has offered itself in London for a great monumental design. With the Tower of London to set the scale, with the splendid waterway it was to span, with all the past and present of the City of London to symbolize, this bridge might have been a monument of the greatness of the British Empire ; and it is—what it is.

London Bridge completes this short survey.[1] Till nearly the middle of the eighteenth century the old bridge was the only bridge possessed by London, and it is amazing that, constructed as it was, it should have lasted for over six hundred years. It was begun in 1176, and appears to have been finished early in the thirteenth century. It was 926 feet by 20 feet wide, and formed with twenty arches, with a drawbridge in the centre. The citizens of London at once proceeded to load it up with houses, and the multiplication of piers so blocked the waterway that the bridge was in constant danger of being washed away by floods. Five arches went in 1282, destroyed by drifting ice. The houses caught fire from time to time, and the passage under the bridge became more and more dangerous. In the aquatint view of London Bridge by Milton, the water is shown rushing through like a millrace. In the middle of the eighteenth century an Act of Parliament was

[1] See Britton and Pugin, *Edifices of London*, ii. 303-13.

obtained for the removal of all buildings on the bridge, and in 1759 Dance, the City architect, and Taylor converted two of the old arches into a large central arch. Finally, some six hundred years after its building, it was decided to remove old London Bridge and build a new one close by. This was before 1801, but it was not till 1824 that the new bridge was actually begun from the designs made by John Rennie and under the superintendence of his son. The *Sunday Times* for November 23, 1828, records that on November 23, 1828, the keystone of the last arch ' was slowly lowered amidst discharge of cannon to its place. The Lord Mayor took a mallet in his hand and struck the stone three times. On the third stroke the whole assembly gave three cheers.' In the Guildhall there is a collection of admirable pencil drawings by E. Cooke, R.A., showing the old bridge in various stages of demolition and the construction of the new.

The bridge is on five arches of unequal spans, instead of the nine of Waterloo Bridge. Rennie increased his span from 120 feet to 150 feet in the central span, a daring piece of construction which J. H. Mansart had foolishly attempted in a bridge over the Allier at Moulins, which totally collapsed within ten years of its being built. Yet London Bridge is disappointing. Instead of keeping his courses the same depth, Rennie, following Mylne, reduced them as they ascended. He omitted any solid walling above the piers, carrying his balustrade through without a break; in both cases with most unfortunate effect. It is probable that these alterations were forced upon

Rennie, or introduced by his son as the work proceeded, as they do not show in the contemporary illustration in Britton and Pugin ;[1] but the design throughout is dull. If one stands on the Old Swan Pier and looks up the river, there is the South-eastern Railway bridge to Cannon Street. It is not lovely, not does it affect to be anything but what it is, a solid, ugly railway bridge ; yet when the tide is running out and the wind is in the sky and the grey water comes swirling under these piers, this bridge, too, has its quality—it sends the imagination roving to other lands and far-distant ages.

So we are back again at the point from which we started, that a great bridge should be something more than a mere means of transit from one side of the river to the other. It has an imaginative significance not lost sight of by the great bridge-builders of the past and still waiting to be recovered. It is not enough to throw a girder across the river or suspend a roadway with steel cables. There is here a chance of architecture on the grandest scale, and such a chance is surely before us, if and when the great Charing Cross scheme is realized. A bridge here is a crying public want on practical grounds, but it carries with it an unequalled opportunity for a fitting memorial *urbi et orbi* of the Great War. Three years ago,[2] with my colleagues, Sir Aston Webb and Mr. John Burns, I put forward a tentative suggestion for the

---

[1] See Britton and Pugin, *Public Edifices* (1828). A parapet wall is shown with breaks over the piers, greatly to the improvement of the design.

[2] In 1916.

line this bridge might follow, having the spire of St. Martin's in the Fields at the end of the vista looking north. On some such lines as that a great scheme might be worked out, with noble approaches and vistas, the best of our sculpture, the finest of our architecture, to speak to future ages of the heroism and sacrifice of the unnumbered dead and of the patient devotion of those men and women who worked at home, without reward, from simple faith and patriotism. Waterloo Bridge, though not built for that purpose, is still the noblest monument of the men of 1815, and in its perfect scale and admirable restraint shows us the way. We have passed through the fiery furnace, and the record of this war shows that the spirit of 1815 lives stronger than ever throughout the Empire. What finer symbol of that spirit could there be than some great bridge such as Waterloo, if only we can rise to the level of our opportunity?

THE TANGLED SKEIN

ART IN ENGLAND, 1800–1920 [1]

THE Hertz Lectures were founded to encourage the study of the arts in relation to civilization, and to-day I shall endeavour to place before you a brief survey of English art from the end of the eighteenth century down to the present day, in the hope that in this way we may be able to clear our ideas in regard to the past, and arrive at some reasonable forecast in regard to the future.

It seems to be generally admitted that there is something wrong with the arts. Are we moving on? Are we only treading a wearisome circle? Or have we reached the final bankruptcy of art, and is there nothing for it but ' tabulae novae ' and the clean sweep? There are those who despair of advance, who point to the mighty men of old and to our own littleness, and maintain that the best we can do is to arrange variations on well-worn themes. There are those again who would simply delete the work of the last four hundred years, and revert point-blank to what they suppose to be a purer art, asking us to take it for granted that the Renaissance, the greatest intellectual adventure of the last thousand years, was itself a mere revival. And there are those who ask us to scrap everything, traditions, associations, all the splendid inheritance of the past, and to paint, model, and design with results unlike anything that ever has been on land or sea. On the other hand, there are some who

[1] The Fifth Annual Lecture on Aspects of Art (Henriette Hertz Trust), delivered before the British Academy 5th May 1920.

believe that the arts, all of which have their origins far back in the past, cannot be violently pulled to pieces and turned upside down without injury to civilization, who think that it is neither necessary nor desirable to seek inspiration in the methods of the South Sea Islanders, and that if the attainment of excellence in the arts is arduous beyond belief, yet, given that price, the arts take their place with other forms of intellectual activity, and are not to be ignored and treated as things of no weight, but are to be recognized as essential features in any reasonable theory of life.

With such divergent views it is well from time to time to take stock of our position. How do we stand after the last hundred years of experiment? We have tried track after track, and again and again found ourselves in a cul-de-sac. Perhaps it was this series of false starts that made William Morris say that we should forget the past and begin again like little children ; but Morris was thinking of architecture and the crafts, and meant that, instead of strutting about in fancy dress, we should think out things for ourselves and find our own methods of expression. His words have been wrested from their meaning, and the attitude of wonder and humility, which as a modern writer insists is at the back of good work in the arts, has been twisted into a licence to parade deliberate childishness and calculated folly as the finished expression of art. The revolutionaries say that the arts all round are in such a hopeless state that nothing short of complete reconstruction can mend them ; and they, at any rate, are making vigorous efforts to carry out their programme by beginning with the extermination of traditional art, both by their own demonstrations, and by well-organized propaganda. One must admit, however, that all is not well with the arts, and that, in regard to their intellectual background, the point of view from which art should be approached, appreciated, and practised, we are worse off in the year 1920 than we were a hundred years ago. We have lost our tradition, and the public has got no standard of its own. The one clear voice of art, once understood of all men, has lost itself in an incredible confusion of tongues, and a general anxiety to shout louder than one's neighbour.

Let us go back to the beginning of the last century. What was the position then in painting, sculpture, and architecture? The giants of the eighteenth century had gone. Richard Wilson died in 1782, Gainsborough in 1787, Reynolds in 1792, and Chambers, the architect, in 1796. When the nineteenth century opened, only one original member of the Royal Academy remained, and that was that successful but moderate artist, Paul Sandby. Benjamin West, ambitious but feeble, was President, and Fuseli had just opened his gallery of illustrations to Milton. Sculpture was represented by Banks, Nollekens, and Bacon, architecture by George Dance the younger, dull, respectable, long past the triumphs of his fortunate youth. The Academy was passing through one of those phases of mediocrity which seem inevitable after periods of great distinction. But among the younger men Hoppner and Lawrence were already Academicians, Shee and Flaxman were elected full members in 1801, and J. M. W. Turner in 1803.

I need not dwell on that memorable chapter in English art, the rise of the English School of landscape painting, the greatest achievement of English art of the nineteenth century. What Gainsborough and Reynolds had done for portrait painting, Turner and Constable did for landscape. These men placed English painting in the front rank of the art of modern Europe. Turner's advance from the classic sobriety of his early manner to the visions of his later work is a development almost without a parallel in the history of painting. Queer, morose, uncouth as he was, Turner was the greatest adventurer in art that this country has ever produced. His career, at least the latter part of it, was a voyage of discovery in regions hitherto uncharted and unknown. If, as an architect, one is tempted to say hard things of the Romantic movement, in modern landscape painting it has been of inestimable value. Without it we might never have had the ' Ulysses and Polyphemus ', and Constable might never have painted ' Weymouth Bay '.

While Turner was storming his way into the mystery of nature, the son of the miller of East Bergholt, with less genius, but more single-minded sincerity of purpose, was mastering her secrets, the cloud and sunshine, the glint in the sky, the ripple on the water, the pageantry of the trees in the countryside that he loved so well. Unequal in execution, sometimes uncertain of vision, Constable never faltered in his resolution to give the exact impression made on him by the nature that he saw. It was this patient loyalty that enabled him to paint a picture so modern in the best sense of the word as ' Weymouth Bay ', with the clouds climbing up the sky from behind the low outline of the downs. Constable painted this picture two years before Turner painted his glorious vision of Ulysses. Differing as they do in almost every point, in idea, in intention, in execution, they have this in common : that each picture, from its own standpoint, gave the death-blow to the classical formula. While Constable showed that nature had something better to show than Sir George Beaumont's brown tree, Turner revealed to a somewhat jaded public the fairyland of pure Romance. Here then was one great achievement of English art, established on such solid foundations that the assaults of the Pre-Raphaelites, some two generations later, scarcely affected it at all. The best of modern landscape art has advanced on the lines laid down by these two great artists, profiting from both, yet never halting in its resolute study of nature, its determination to see with its own eyes, and express itself in its own terms. The latest methods of landscape painting, having no relation to realities, I may leave over for the present.

In other branches of painting the development has been less remarkable. The tradition of Gainsborough and Reynolds was carried on by Hoppner and Lawrence, Shee, Raeburn, Watts, and Millais, and in spite of recent vagaries and eccentricities is maintained by the best of our painters to-day. Nor was there in the nineteenth century any radical change of outlook in genre and figure subjects, except for the strange interlude of the Pre-Raphaelites. The mantle of Hogarth was too heavy for his successors, but Wilkie and Webster, Leslie and Edmund Ward, Frith, too, in his own way, though their reputation has greatly diminished, were considerable men in their time. Their weakness was that they concentrated their effort on exactness of anecdote rather than on the artistic opportunity of their subject. The 'Surrender of Breda' is one of the great pictures of the world, not because it is an exact historical document—it is probably quite inexact—but because it is a magnificent vision ranging far outside the limits of what may have been a mere prosaic incident, summing up, as it were, the pageantry and historic greatness and hardness of Spain. The Pre-Raphaelite movement, highly interesting as it was, seems to me to have been a curious episode in the history of English art, unaccountable except as a by-blow of the Romantic movement, galvanized into a brief but strenuous life by the cant and commercialism of the middle period of the nineteenth century. It failed of its purpose in painting because it was not so much an artistic movement as an ethical revolt against conventionality, and when artistic movements get entangled with ethics and politics their collapse is inevitable, because they confound things which have no common denominator. The Pre-Raphaelite movement, as a movement, fell to pieces almost at once. On its artistic side it was inspired by a rather narrow view of what nature really is. The 'Flight into Egypt' or the 'Ophelia' in the National Gallery are not in fact more faithful to nature than Constable's 'Weymouth Bay', if as much so. The Pre-Raphaelites seem to have thought that fidelity to nature consisted of peering at natural objects through a microscope, and reproducing what they saw in the minutest detail. Yet in point of fact the eye never takes in or sees a tithe of all the detail laboriously collected by the Pre-Raphaelites, and effects of light and shade and atmosphere, of colour, values, and tones, are equally part of the visible appearances of nature, not less deserving of study and record, not less insistent in their appeal to the emotions, than the blade of grass or the shavings in the carpenter's shop. One has to discount the tendency of poets and artists to pose and do strange things, but the Pre-Raphaelites were sincere in their revolt against convention, and it is not easy to reconcile the intensity of feeling of 'The Carpenter's shop' with the later work of the great painter who had produced that wonderful picture. The Band of Brothers was broken up, and their influence disappeared, but some of our younger men seem again to be moving in that direction, and this and the rigorous discipline it implies may yet save English art, and one ought to recognize that, though the Pre-Raphaelites failed in painting, their revolt against the deplorable art of 1851 was successful in another direction. Through William Morris they started a movement of far-reaching importance in the reorganization of the Arts and Crafts.

Now let us turn to sculpture and architecture. There can be no getting over the fact that in all the three arts England was later in arriving than Italy and France. In painting we produced no considerable painters till the middle of the eighteenth century. Isaac Oliver, the miniaturist, was probably a Frenchman. Lely was born in Westphalia, Kneller in Lubeck ; and it was worse in sculpture, because at one time we had possessed, in our English imagers, artists who could hold their own with any one of their time. I know no finer figure of its kind than that recumbent warrior of the fourteenth century on an alien tomb at Tewkesbury ; or, a hundred years later, the knight in alabaster in Berrington Church, tall, limber, and athletic, the vivid portrait of some young gallant ' fato sibi immaturo suis acerbissimo ereptus '.[1] But this school of sculpture disappeared. The tomb-makers of Elizabeth and James the First drew their inspiration from the Low Countries, not from France or Italy. Old Nicholas Stone, the faithful friend of Inigo Jones, and the earlier seventeenth-century sculptors in England were nowhere near the level of Goujon, or Germain Pilon, or even of their successors, the Biards, Guillains, and Anguiers ; and though Grinling Gibbons and his school were dexterous carvers and ornamentalists we have simply nothing to show that is comparable to the work of Coysevox and Girardon, Pierre Puget, and Robert le Lorrain. Had it been possible for William III to bring over with him artists such as these, the whole course of monumental sculpture in this country might have been altered. We might have had, what we never have had in England, a tradition of great monumental sculpture. Our men did well enough with minor mural monuments, and head-

[1] Inscription on the monument to Edward Valentine Blomfield in the Cloisters of Emmanuel College, Cambridge— younger brother of C. J. Blomfield, Bishop of London, a Fellow of Emmanuel who died in 1816 at the age of 28, after a very brilliant career at Cambridge.'

stones, and the vases and lead figures of our older gardens ; but for important work our ancestors at the end of the eighteenth century had to be content with the classicism of Banks and the futilities of Nollekens. These men were capable technicians, John Bacon the sculptor in particular, but they followed blindly the classical convention of the time, and one finds in the lists of the Academy a dull succession of famous men : Flaxman elected an R.A. in 1801, Rossi 1803, Westmacott in 1815, Chantrey in 1819, Baily in 1822, and in 1837 an artist who posed as the protagonist of the antique but was in fact a master of clap-trap, John Gibson, the sculptor of the tinted Venus. Of all these men Flaxman is the most interesting, and the best known through his outline drawings ; but, attractive as one used to find these drawings years ago, they are not convincing. The figures strike heroic attitudes which narrowly escape being absurd, and Flaxman's line, instead of being subtle and flexible, clinging as it were to the idea, is fixed and conventional, seldom closely related to the physical facts of nature. None of these men could found a school or establish a tradition. They pursued, at second hand, a false ideal of classical art, inspired by Winckelmann and realized by Canova. They ignored the differences of race and temperament which separated their countrymen from the ancient Greeks, and though they enjoyed considerable success (Nollekens left £200,000) because their patrons accepted what was offered them as being in the movement, they nearly ruined sculpture in England. The art lost its interest for the man in the street. It was so remote and academic that its very existence was almost forgotten. Yet the game was not wholly lost. Our national monuments of fifty years ago are not worse than the stumpy little figures by David of Angers which disfigure most of the big provincial towns of France. John Bell's Crimean memorial is a grave and dignified work, far better than many that have followed it ; and then there is Alfred Stevens, one of those rare men of genius that it is given to this country from time to time to produce, men who flash through the sky like a meteor, out of the unknown and back to the unknown, men not to be accounted for, whose bow is too great for other men to pull. Yet the tragedy of the Wellington Monument shows how far the sentimentalist and the rhetorician had misled the public. While Ruskin's lecture room at Oxford was crowded with enthusiastic young ladies, Alfred Stevens was breaking his heart over the Wellington Monument, the greatest masterpiece in the whole range of English art, and no helping hand was ever held out to him by the Slade Professor. Stevens gathered round him a few able and devoted men, but he founded no permanent school, and his influence over the general average of modern work has been less than might have been expected. The sense of great monumental design, the power of thinking in terms of architecture and sculpture in complete harmony, is still rare in this country. Splendid work is being done by our sculptors to-day, but it is as individuals, each in his own way. There seems to be even less common ground among them than there is among architects. One recognizes at once French work or

Italian work, but, brilliant as they are individually, our sculptors have formed no school. Perhaps it is the penalty of our inveterate individualism. Possibly we shall never be able to establish a school of sculpture such as Colbert built up at Versailles ; or a tradition such as that which began with Jean Goujon and only failed in the closing years of the eighteenth century, when most of what was most admirable in French art perished with it.

The course of architecture has been different. At the beginning of the eighteenth century there existed in England the finest tradition of technique ever possessed since the days of the fourteenth century. A hundred years later we had lost that tradition. Architecture showed signs of splitting up into several antagonistic groups. Chambers, 'Ultimus Romanorum', was the last of the Palladians. Robert Adam had swept the board with his fashionable adaptation of Roman ornament. Here were already two divergent manners of design. The Neo-Greek movement was in the air, and the vague sentimentalism of the eighteenth century was already gathering its forces to crystallize into the Gothic Revival. It was, too, one of the misfortunes of the latter part of the eighteenth century, a misfortune which has since developed in alarming proportions, that, in the disintegration of the classical tradition, the arts became a fashion. The layman not only began to interest himself in art, but took on himself to lay down the law about it. The Comte de Caylus in France, and Horace Walpole in England, did an infinity of harm by taking the arts under their patronage, and by using their social position to influence opinion in a definite direction, and the worst of it was that the artists weakly acquiesced, and lent themselves to the whims and caprices of their noble patrons. The old French Academy was suppressed in 1793, and in England the nineteenth century has little to show in architecture but a succession of experiments in every conceivable manner of Western architecture, experiments which seem to us now all equally stale and futile, because they never went to the root of the matter, and failed to recognize that architecture is the art of ordered building. We turn again to the lists of the Academicians, and find that John Soane was elected an R.A. in 1803. Soane, a man of enthusiasm and ability, belonged to that formidable class of artists who, with an extreme desire for originality, possess no instinctive feeling for beauty and lack the restraint of a fastidious taste. Being possessed of much energy and self-confidence, he impressed the public with his own belief in himself, and led it seriously astray by his misplaced attempts at original design, so that when Smirke became merely dull, Soane was sometimes positively vulgar. The century opened badly in architecture, and the scholarship and refinement of Wilkins and Decimus Burton, and later of Cockerell, could do little to stay the descent into the quagmires of archaeology. Architects and their public were losing their way, bewildered first by copies of Roman architecture, then by copies of Greek, in both cases regardless of the real purpose of architecture. With the barriers of tradition withdrawn, any flotsam and jetsam might come floating down the stream, and why not Neo-Gothic as well as Neo-Greek or Neo-Roman ? In these troubled waters the Romanticist and the Pietist saw their chance. Medieval architecture was collected and definitely labelled as the only possible expression of the fervent soul and the Christian religion, and any other form of architecture was excommunicated as essentially immoral. With this introduction of ethics, bad history, and controversial elements irrelevant to the art, the real purpose of architecture as the art of building was forgotten ; the mere shell, the trappings and accessories of one particular style were set up as axioms of architecture, to deviate from which was to commit a deadly sin. It did not occur to these fanatics that the words of a language are nothing; it is their use in expressing ideas that makes the difference between the poet and the costermonger. But unfortunately for architecture, the protagonist of Neo-Gothic was a writer who combined an extraordinary gift of eloquence with an entire absence of critical sense of architecture, and though the ' noble patron ' was bad enough, the amateurs of architecture in the last century were a thousand times worse, because they criticized the art on irrelevant grounds, and it is to these, and to those professional men who aided and abetted them, that we owe our tale of woes ; the spoliation of our churches and cathedrals under the guise of restoration, the substitution of pastiche for design, the pursuit of the picturesque, the affectations of modes of thought and expression that have long ceased to bear any relation to life, culminating in the literal reproduction of old buildings and even in their bodily removal to America for re-erection.

Folly and false sentiment could hardly go further, and it has led to the inevitable consequence, that architecture has lost its place of honour. People ceased to take it seriously; indeed within the last few years one writer stated that it was just an affair of fashion, like the latest hat from Paris. Another had a short way with artists, and settled it once for all by asserting that an artist was like a pastry-cook, all he had to do was to please the person who bought his pictures or ate his pies; and another writer has made the remarkable discovery that the criterion of art is whether the artist does or does not take pleasure in his work, a formula of much encouragement to the hosts of the amateur, and our latest and most violent critic asks us why we do not join the army of Bolshevists and blow up the whole business. To such a pass have we been brought by the sentimentality of the nineteenth century, and the muddle-headed thought of this.

There can be no doubt that things were better a hundred years ago. It is not that we lack good architects, painters, and sculptors, but the body politic of the arts is not sound. There were bad artists in the eighteenth century, but not a tithe of what there are in the twentieth. Bad work was done, but it was seldom accepted as good, whereas nowadays it is advertised as the last word of genius. There existed in those days a definite standard of technique and craftsmanship, now there is no such standard. Artists themselves seem uncertain in their ideals, lacking in confidence and settled purpose. The experiments of the nineteenth century in arts have not been happy, and their mischievous effects have been intensified by the well-meant efforts of the State, still inspired, it seems, by the ideas of the Great Exhibition, ignorant of art and at heart indifferent to it. As usual in England, art has been translated into terms of politics. Our State educational authorities seem to think that artists can be turned out by the gross, given the necessary State-aided machinery, and do not realize that a result of their benevolence has been to set a premium on incompetence.

The results of our retrospect are not very encouraging, but the worst foes of modern art are those of our own household, or rather the strangers within our gates. It is the old story of the ivy that kills the tree it grows on; and it is an old story in another way, for it is probable that artists gave themselves away when they started exhibitions in the eighteenth century, and unwittingly provided occasion for the whole dynasty of critics, with Diderot, brilliant and witty, as the first of the line. The critics have not been slow to discover that it is easier to write about the arts when the flow of eloquence is not impeded by knowledge of the subject or any acquaintance with their practice; and they have in recent years advanced to still further heights by inventing an Aunt Sally of what they call 'Academic Art', and rigging up a fantastic theory of aesthetic out of the studio talk of the revolutionaries. Studio talk is often stimulating, always amusing—so in our undergraduate days we used to bandy our crude theories of philosophy. But we were content to leave it at that, whereas in the arts our teachers seem to dread being left behind and so are constantly on the look out for the latest theory. In the arts also it seems to be the fact that those who are most prolific of theories are often the least capable of carrying them out, and find it necessary to substitute the written or spoken word for the legitimate methods of expression of their art, and to cover up their technical shortcomings by the invention of a series of formulas which have this in common, that they one and all dispense with technique. Thus a movement which may have begun with a genuine desire to extend the traditional limits of art has ended in an ever-accelerating rush for patent medicines. No sooner has one formula attained notoriety than it is superseded by another, and we now look for a new school every season. The impressionist, the luminist, the pointillist, and the cubist are already old-fashioned. From the literary point of view there is more incident and material in these desperate scrambles than is to be found in the work of the sincere and patient artist. 'Épater le bourgeois' at all costs, is the war-cry of the new art, and after all papers must sell. Hinc illae lacrimae, and the lucubrations of our art-critics. Like Molière's 'gens de qualité' the art critics 'savent tout sans avoir rien appris'; and thus, without technical knowledge of painting, sculpture, or architecture, they are able to instruct us what we are to admire in art, what is the business of the artist, and how he ought to carry it out. As for the artist, he is to have no voice in the matter at all, he must just do what he is bid, or take the consequences of being left out in the cold. It is true that this makes the position of the artist even worse than the βάναυσος of Aristotle. He, though a servile person, was at least allowed to do his own work his own way. It is not believed that Ictinus was seriously disturbed by the bawlings of Cleon, whereas if the sculptor or painter of to-day disregards the nod of Jupiter Tonans he is done for, down and out.

The plain man may well ask what it is all about, what it is that these gentlemen who lecture us daily and weekly really want us to think? It is at this point that our teachers fail us. In fact, they seem to contradict each other, and though they speak with much apparent intensity and even bitterness of feeling, the substance of what they want is so elusive as to make one sometimes doubt if there is any substance at all. A little while ago our papers announced to us the presence in a London gallery of a picture which the critics assured us transcended all contemporary art, and this was followed up by an exhibition of the work of a well-known, or I must be permitted to say notorious, French painter. The critics as one body rose to lyrical heights in their raptures. We were told that this was no mere presentation of life, but life itself; a revelation, as it were, of some quintessential mystery of existence. We went in the requisite spirit of humility to that exhibition, and what did we find? I do not want to use harsh words, and will content myself with saying that one found a collection of canvases that appeared to have no meaning at all and no object, except the negation of every quality of form, colour, and composition that painters in the past have ever sought to realize. Referring to a canvas at another exhibition, a critic, in calling for our admiration, hit upon the elegant formula that the artist ' had not allowed nature to intrude her irrelevancies'; and indeed he had not, for no ordinary person looking at this paint and canvas could have formed any idea as to what it meant. And this is the case with nearly all this work. As it stands it is unintelligible, and sometimes it is difficult to escape the impression that it is intended to be so. Yet we are constantly assured that this is the last word in art, and that any other kind of art is stupid, sentimental, squalid, vulgar, banal, meaningless, and many other opprobrious terms, all of which may be found in the latest exposition of the new gospel of art. Its apostles do not always agree in their doctrine, but after a careful collation of their opinions I have come to the conclusion that, by incessant juggling with words and industrious spinning of spiders' webs, they have landed themselves in a position which is out of relation to art altogether, unless by art is meant something different from what so far in the history of civilization it has always been taken to mean. Nature, that is to say, the visible appearances of nature, are ruled out. 'It does not matter', so runs the latest revelation, 'what objective nature supplies. The inventive artist is his own purveyor.' Like the spider he spins out of himself. The text says, 'The best half-dozen artists of any country, as regards the actual beauty and significance of their work, do not depend on the objective world for their success or stimulus'; which indeed would be perfectly true in the sense that the ore does not become silver and gold till it is transmuted, but which is quite untrue when it means that the artist is independent of the visible appearances of nature.

Thus another writer says : ' They use their paint for the reconstruction of their impression rather than for the representation of what they have seen ', a formula which describes with some precision the efforts of the savage, or the child with its first box of paints. A painter who by the mercy of Providence has not yet penetrated to England is referred to with approval as ' the first painter to make pictures of purely non-representative forms and colours ' (by which we must understand form and colour which have no resemblance to any natural object whatever) ; and this remarkable exploit is introduced to us as ' expressionism '. Nature, the essential model and material of expression in all great painting, is henceforward to be regarded as out of date. She may furnish suggestions on which the genius of the artist may browse and ruminate, but after that, or even without it, the artist is to turn his back on her ; to borrow the phrase of an acute and thoughtful writer in the *London Mercury*, he is to be relieved ' altogether of the irrelevant incubus of representation '. Indeed, the logical conclusion would be that he should shut his eyes entirely, lest nature should contaminate the spiritual purity of his vision ; for all is done, as children would say, ' out of his head '. He might just as well do it on his head, so far as the spectator is concerned, for the artist's concern is solely with his own emotions, and if the result has no meaning for anybody else, that is their fault, and no affair of the artist. I gather that it is unnecessary for him to be intelligible. Art is to be an affair of hieroglyphics, of arrangements of forms and colours which are out of relation to observed realities and indeed which need have no meaning at all, because there are always at hand the skilled art critic to supply the necessary hermeneutics, and the more unintelligible the artist, the better material for the eloquence and ingenuity of the critic. Judging by the results, no training would seem to be necessary; all one has to do is to learn how to mix a few colours, draw any old line, and splash some paint on the canvas. Indeed it is a favourite position of the new school that imperfect technique is a necessary condition of all great art. ' All great works of art ', we are told, ' show an effort, a roughness, an inadequacy of craftsmanship which is the essence of their beauty.' Cimabue is a greater artist than Michael Angelo, and Matisse, I suppose, than Turner, and by the acceptance of this simple formula artists henceforward will be freed from the necessity of spending long laborious days in the study of their art. All they have to do is to feel ferociously, and as there will be no standard but the emotionalism of the critic, and no criterion but the potentiality of literary copy, the quality of feeling is immaterial. I am grateful to the erratic author of a recent pamphlet for coming out into the open, and boldly stating a position which so far has failed to reveal itself through the clouds of words under which the critics conceal their batteries ; and it comes to this, that art is no longer what the great masters have always taken it to be, the trained power of expressing in terms of form and colour and ordered design ideas which can be given their fullest and best expression only in those terms, but is a bedlam business in which you say anything you like in any way you please. The only thing one objects to is that these theorists will insist on calling this ' art '.

The writer to whom I have just referred has the greatest contempt for the humble architect, and he asks us, ' Where is our vortex ? ' I do not affect to know what this means, but there is the familiar answer in the *Clouds* of Aristophanes. ' What a splendid thing is learning ! ' says Strepsiades. ' Zeus, my Pheidippides, exists no longer.' ' Who does exist then ? ' asks his son, and the answer is

$$\Delta\hat{\iota}\nu os\ \beta a\sigma\iota\lambda\epsilon\acute{v}\epsilon\iota, \tau\grave{o}\nu\ \Delta\acute{\iota}'\ \grave{\epsilon}\xi\epsilon\lambda\eta\lambda a\kappa\acute{\omega}s,$$

' Vortex (chaos) is king, having dethroned Zeus.' Adolescens Leo, as Matthew Arnold used to call him, has won the day, he has thrust out the artist and sits triumphant in his place. One had hoped that the grip of reality brought about by the war would have cleared this rubbish away ; but at present the last state of this man is worse than the first. The seven devils have entered in and rallied their forces for a final assault on our sanity. In the presence of certain recent works held up to us as admirable art, one may well ask, ' Am I off my head, or is the man who tells me this ? ' It is time that a halt was called in this race for the lunatic asylum. ' Non tali auxilio nec defensoribus istis ' will art advance. The priests of Baal may gash themselves with knives, but the Lord will not hearken. The old road is still the only road, and there is still but one way for the artist, the unwearied effort to perfect his power of expression in his art, the patient study of colour, of light and shade, of form and its ordering, thought and invention, and the sure hold of the artist's own ideals, no matter what the critics say or the fashion calls for. Only so can the artist qualify himself for his high calling.

An ingenious writer in some recent essays on art appears to regard the artist as a mere purveyor of thrills, and makes the sensibility of the critic the touchstone of criticism. I would remind him that the artist, after all, has a soul, and that one of the compensations of his arduous life is that he has a world of his own, to which he alone has the key, and into which the most brilliant critic can never enter. Art criticism, after all, is not the final cause of art, and if the critics would only leave the artist and his work alone for a time there would be a chance of the re-establishment of a wise judgement in these things, and meanwhile the critics might profitably employ their time in the serious study of the history of art, and refrain from endeavouring to screw down its practice to their speculations in metaphysics. It would be charitable to find the motive of the new gospels in a sincere desire to reach the secret of aesthetic enjoyment, but it is difficult to get rid of an ever-recurring suspicion of charlatanism.

Our hope lies with artists themselves. Fresh problems are constantly rising. We cannot stand still, and it is a good sign that, among artists, there is no disposition to do so. Even our revolutionaries may go far, if only their heads can be set in the right direction. The folly of modern art is due to the camp-followers rather than to artists, for there is among artists to-day a spirit of adventure, which must be the driving force behind any advance. Even the despised architects, at least the best of them, have ceased to treat architecture as an affair of sketch-book design. We no longer quarrel about Classic and Romantic because we realize that there is no necessary conflict between them, and that in all fine work there must be elements of both, and through much toil and tribulation we have caught some insight into the real meaning of the classical spirit, a mere glimpse, it may be, of what was at the back of the minds of the men who designed the temples of Egypt, the Parthenon, the Doric of Paestum, the Colosseum, or the Porta del Palio. True Classic in architecture means clarity and simplicity, the elimination of the unessential, the absolute statement of the purpose of the design, including in that purpose the whole range of appeal to our imagination and emotions. M. Lanson recently defined the tendency of contemporary French literature as ' a broad-minded classicism which will not display hostility towards romanticism and symbolism, for it will embody in its aestheticism and its creations the best results of the literary activities of the nineteenth century '. So in the arts it is our business to profit by the labours of those who have gone before us, but it is also our business to set our face forward, to throw overboard all that is meaningless, and to search out our own expression for our own problems and ideals. This, I believe, is what the best of our men are doing, whether they are young or old. But owing to the fact that artists are much more rare than is generally supposed, the drift of this effort is missed. Yet each new enterprise in the arts is surely a voyage of discovery, a raid on latent beauty, unthought of by the profane, and in the first instance only seen afar off by the artist himself. It was this impulse that inspired the Renaissance, no mere pedantry, no relapse into paganism, but the spirit of high adventure in the whole range of thought, casting backwards and forwards ; questing sometimes in strange lands, yet ever intent on great ideals never to be fully realized.

It is in the same spirit to-day that the hope of the future lies ; the spirit of adventure, steeled by discipline, and armed with full knowledge of the resources of art. For beauty is not to be snatched at or caught by chance, or won by the tricks of the mountebank. The suitors in Ithaca perished miserably. It was Ulysses, the man of much daring and many wiles, who, after years of strenuous labour and perilous adventure, won back his kingdom and his queen.

GREEK ARCHITECTURE

NOBODY has ever disputed the beauty of Greek architecture. We admit the justice of a description of the Parthenon as ' le suprême effort de génie à la poursuite du beau ' ; but the layman must sometimes ask himself what does it mean ? Where did it come from, where did it go to, why is it thought so beautiful, how was it that this people, relatively insignificant in power, in territory, and in numbers, was able to attain to this astonishing supremacy in art. These are questions not easily answered. The evidence is fragmentary and not always conclusive, the ruins of a few temples and buildings, a technical treatise by a garrulous third-rate writer in the first century A. D.,[1] the anecdotes of an indefatigable collector[2] a little later, the notes of a traveller in the second century[3] and the materials collected by the patient research of scholars and archaeologists, pieced together on more or less ingenious hypotheses. Indeed, a great part of what is written on Greek architecture is simple hypothesis. There is not much to go on, yet Greek architecture (and by this I mean the architecture of the sixth and fifth centuries B.C.) remains one of the great outstanding facts in the history of the architecture of the Western world, and the art of the age of Pericles is the fountain-head to which artists still return.

[1] Vitruvius, *De Architectura*.
[2] Pliny the Elder, *Historia Naturalis*, xxxvi.
[3] Pausanias, Περιήγησις τῆς Ἑλλάδος.

Where that art sprang from and how it grew is largely a matter of speculation. There have been legends of civilizations wiped out in tremendous cataclysms that left no trace behind them. Vague suggestions are made that the cradle of the race was in Asia. All we know for certain is that the earliest civilizations of which actual historical evidence remains are those of Chaldea and Egypt, and that the art of these countries reached a high degree of attainment long before we come upon the earliest traces of art of any sort in Greece. That both these countries contributed in varying degrees to the art of Greece is certain, but that is not the whole of the story. As we shall see, another element comes into play, which made of that art almost a new creation, differing in outlook and ideal from any art that preceded it, stamped by the genius of a vigorous northern race with a character all its own. The art of the East and the art of the West never really fused. There is a difference in kind between the joyous vitality of pure Greek art and the gloomy vision of Asia, with its craving for the vast and terrible, its sombre imagination, its lack of humanity and indifference to the individual.

It is not, however, till far down in the progress of history that this differentiation asserts itself. Greek art is relatively a late development. The Great Pyramid at Ghizeh was built some 2,000 years before a stone was laid of the masonry of Mycenae. The Hall of Columns of Karnak, with its columns sixty feet high, was probably coeval with the Treasury of Atreus : in other words, when the art of Greece and of the islands was scarcely out of the barbaric stage, a wonderful art had been in existence across the Mediterranean from time immemorial. Both Egypt and Chaldea attained a high degree of civilization long before the Dorians were ever heard of. At some remote period the Egyptian influence penetrated to Crete and Cyprus, the islands of the Aegean and the mainland of Greece ; and the intermediaries were the Phoenicians, that enterprising race of merchant adventurers, whose home was in Syria, and whose fleets traversed the Mediterranean from east to west. The Phoenicians were traders and not artists. In Egypt they came into contact with a highly developed art, beyond their comprehension in its essential features, yet including details which could easily be apprehended by their quick commercial intelligence. Wherever they touched on their voyages, Cyprus, Crete, the southern islands of the Aegean, the mainland of Greece, the south of Italy, Sicily, Carthage, the Balearic islands, and Spain in the Far West, they probably carried with them, for trading purposes, minor articles of Egyptian workmanship which may have supplied hints to the indigenous peoples.[1] Where they established settlements, they reproduced what they could recollect of the methods of

[1] In a hypogeum discovered at Byblos in Syria 1923 M. Virolleaud found a limestone sarcophagus 2.80 m. long × 1.48 m. × 2.32 m. high ; also near it a good deal of worked sheet gold, and an Egyptian perfume vase of obsidian with gold circles at neck and foot, and gold pieces fitting spaces on the top ' bearing in bold relief the hieroglyphic signs which make the name of Maātenrā, the enthronement name of Amenemhat III of the 12th Dynasty ' (about 2,000 years B.C.). *Times*, 21st May 1923.

Egyptian architecture, possessing at second hand a knowledge of technical methods in advance of anything within the knowledge of the people among whom they settled. Rudimentary anticipations of the Ionic volute are found in Phoenician capitals, vague reminiscences of what the traders had seen in Egypt and elsewhere. Moreover, the Phoenicians, who possessed the skill of sailors in the use of tackle, would have had little difficulty in handling large stones set dry in more or less regular courses, which was a characteristic feature of Cretan and Mycenaean building. It is too soon to describe the work as architecture. It is doubtful if the Phoenicians possessed any aptitude for the arts. Their role was that of intermediaries only.

Obscure as was the part played by the Phoenicians in the early origins of art in Greece and the islands, there was another channel through which Eastern influences came to bear on its development, which is even more uncertain. To the west of Chaldea and north of Syria dwelt a race of which little is known, the Hittites.

Carchemish, their capital, was on the upper Euphrates north-east of Antioch, and their power appears to have extended westward through Asia Minor to the shores of the Aegean. Dr. Sayce says that in the thirteenth century B.C. it extended from ' the banks of the Euphrates to the shores of the Aegean, including both the cultured Semites of Syria and the rude barbarians of the Greek Seas '; he even says that the Hittites ' brought the civilization of the East to the barbarous tribes of the distant West '. What actually remains of Hittite art hardly bears out this statement. When the Hittite power was at its height, Minoan ' art ' had long been practised in Crete, and according to the most popular chronology had already passed its prime and given way to the art of Mycenae and Tiryns. The scanty evidence of Hittite art consists of bas-reliefs of figures and animals cut on the face of rocks along the natural caravan routes through Asia Minor from east to west. This and the evidence of seals and engraved gems show that Hittite art was derived first from Chaldea, later from Egypt. It undoubtedly exercised some influence on the art of the early Greek settlers on the eastern side of the Aegean, and gave it an Asiatic cast, which it never lost throughout all its later developments. For the Greeks of Asia Minor never really understood the austere ideal of Doric art. Ionian art crossed westward to Greece, but the Dorian never went east. It was the art of a strong northern race, that found no place for itself among the softer peoples of Asia Minor.

At this point we can take up the first rudimentary beginnings of Greek art. The discoveries of the last forty years have proved the existence, in Crete and Cyprus, southern Greece, and the islands of the Aegean, of an archaic art of obscure origin, of very great interest and of remarkable attainment in certain directions, long before the earliest beginnings of what we mean when we speak of Greek architecture. So far as architecture is concerned, this archaic art is of relatively minor importance. It plays a small part, if any, in subsequent developments, and though enthusiastic explorers claim to find in it anticipations of the details of modern domestic architecture, the evidence produced is unconvincing. Great movements in the arts always owe some debt to the periods that have preceded them, but Minoan and Mycenaean art, at any rate in regard to architecture, was rather the last word of a decaying civilization than the first herald of the glorious art of Greece in the sixth and fifth centuries B.C. We are still far back in remote ages, remote that is so far as Greek art is concerned, anywhere between 2000 and 1000 B.C. or even earlier,[1] back in the Minoan age of Crete, with its rudimentary architecture and its relatively high excellence in the crafts ; and in the age of Mycenae and Tiryns, the age that produced the Lion Gate at Mycenae, and that strange half-barbaric work, if I may be pardoned the term, the Treasury of Atreus. It is worth pausing to consider these archaic buildings, not so much to show any relationship to later work (which scarcely existed) as to call attention to the fact that the Minoan and Mycenean builders were moving in a direction that would never have led to the column and lintel architecture of the seventh and sixth centuries B.C. It might have led to some form of dome construction, it could never have led to the Doric of the Sicilian temples. No stronger evidence of the genius of the Dorian invaders could be produced than that, with this unpro-

[1] Sir Arthur Evans has drawn up an ingenious chronology of early Minoan (2800–2200 B.C.), Middle Minoan (2200–1700 B.C.), and late Minoan (1700–1200). The evidence is almost entirely that of pottery discovered on the site. The whole question of the relations of Minoan to Mycenean art, and of this archaic art to the older civilizations of Egypt and Chaldea, is very obscure and uncertain.

mising art in possession, they were yet able in the course of three or four centuries to create Greek architecture. The design of the Lion Gate is a strange jumble of ill-adjusted motives. It is set in a wall of great stones roughly squared and laid dry. Two monolith jambs support a huge lintel, cambered in the middle like the tie-beams of our sixteenth-century roofs. Above the lintel the courses are gathered over, leaving between their lower faces and the top of the lintel a triangular space of a steep pitch (about 60°) in which was inserted a frontispiece carved on a single stone, representing two lions standing up on either side of an archaic column, supporting a fragment of a rudimentary architrave.[1] The heraldic pose of the lions and the technique of their sculpture, so suggestive of Assyrian reliefs with their splendid sense of muscular form and energy, are far ahead of an architecture that is still barbaric, scarcely architecture at all. There is here nothing to suggest the Doric of Paestum and Selinus, much to recall the megalithic buildings of Syria and the sculpture of the Further East.

The Treasury of Atreus is still more remarkable, not only because it shows more skill in building but because its design is based on a structural motive which seems to have been wholly abandoned by the successors of the Mycenaean builders. The Treasury of Atreus (or Tomb of Agamemnon) was excavated in a hill, and consists of a long passage about 120 × 21 feet

[1] The heraldic treatment of the lions is of Eastern origin. The Greeks had a tradition that the chieftains of Mycenae came from Lydia.

wide, with retaining walls of megalithic masonry on either side, terminating in a great entrance doorway. This doorway is flanked on either side by columns tapering downwards and decorated with chevrons in a manner very similar to Norman work of the eleventh century, and apparently intended solely for ornament.[1] The entrance opened into a circular domed chamber about 48 ft. 6 in. in diameter, 45 ft. 4 in. high, out of which opened another smaller chamber. The dome, in section, is built on the curve of a parabola, formed with courses projecting over one another, and not set out radial to the curve of the dome. In other words it is not a true dome or arch, but a succession of corbels. The internal face of the dome is dressed down, and was covered with ornament of some sort, whether metal rosettes, or enamelled terracotta, or wholly in metal, possibly the famous gold of Mycenae, is not known. The whole of this chamber was covered in with a mound of earth, in accordance with the primitive custom of concealing the chieftain's grave. It is impossible to find, in this extremely interesting monument or in the domed chamber of Orchomenos in Boeotia, any trace of future developments in Greek architecture. Both in intention and in its psychological background it seems almost as remote from the Doric Temple as the Great Pyramid itself. In point of fact architecture was still in a rudimentary stage. It has been proved abundantly that architecture comes late in the sequence of the Arts. People could draw well long before they could design. Among the cavemen, for example, there were

[1] These columns are now in the British Museum.

GREEK ARCHITECTURE

admirable draughtsmen, but they had to make their drawings on the sides of caves. That there existed in the Minoan and Mycenaean ages skilful potters and metal-workers is shown by the vases of Knossos and such examples as the gold cup of Vaphio near Sparta—that they built habitable buildings and decorated them to the best of their ability is also proved, as, for example, the palace of Tiryns; but it has not yet been shown that their builders reached the degree of skilled design at which building becomes architecture. Architecture had not yet found itself in Greece.

Then somewhere about 1000 B.C. came the Dorian invasions, and the art of Crete and Mycenae vanished into space. Possibly the legend was right which said that the conquered people of the mainland carried it away with them to Asia—the three or four centuries following the Doric invasions are a blank which future research may fill out for us. Anyhow, so far as Art is concerned, there appears to have been a *détente*, during which the new race was settling down to its conquest, finding itself, possibly assimilating something at any rate of the older civilization. The survival of such buildings as the Treasury of Atreus show that the Dorians were not simple barbarians, destroying all that came in their way. Even Sparta in its earlier days was not a mere military machine; discoveries made in 1909–16 suggest that from the ninth to the seventh centuries B.C. Sparta had some sort of an art of its own showing traces of Asiatic influence in its pottery, and a little later Sparta concluded an alliance with Croesus, King of Lydia, and Bathycles, an artist of Magnesia in

Ionia, was treated with honour in Sparta. The Dorians were something more than fighters, they seem to have possessed some civilization, and to have been endowed with a sort of natural capacity for the Arts, which after two or three centuries of experiment will find its own splendid expression within very definite and original lines.[1] The legend of the return of the Heracleidae was to be justified by their later history. No merely imitative race could have evolved the perfect manner of the great Doric temples from reminiscences of Egypt and the East, and the rudimentary buildings of Crete and Mycenae.

Greek architecture for the purpose of this study is Dorian architecture and its elements are simple. It was evolved in the design of their temples, and with the exception of their theatres it was summed up in these temples. From the period during which Greek architecture was being built up to its maturity, say from the seventh century B.C. to the completion of the Parthenon in the fifth century B.C., the whole life of the Greek was coloured and dominated by his religion and its observances; and his religion was not the sinister mystery of Egypt, but on the whole a cheerful open-air Pantheism that gloried in the life and beauty of the visible world in which he lived. He himself was content to live in a poor house, so long as he had his market-place, his ceremonial theatre, and the glorious temples of his gods. Moreover, to whatever depths the Athenians may have sunk in the time of St. Paul, in the heroic days of Pericles they were remarkable for constancy of purpose and the steadfastness of their ideals. They

[1] See note at end of chapter.

stood on the ancient ways, and it never occurred to them to abandon the tradition of their fathers ; their business was to carry it forward to perfection. The result was that the architecture of their temples proceeded on lines that long use had made sacrosanct ; and its technique is summed up in the history of two orders, the Doric and the Ionic.[1]

Now the order, its character, dimensions, and disposition, with the wall of the cella (or enclosed shrine) within the colonnade, summed up the elements, the vocabulary if one may so put it, of Greek architecture ; and we come here at the outset on a curious quality of the Greek genius, and one that differentiates it from the Roman. The properties of wood and stone as materials are clearly different, things can be done with the one which are impossible with the other ; but the Greeks either did not realize this or did not trouble their heads about it. They found that the post and lintel was a simple means of building and they adopted it as their permanent method of construction. If the span became too wide, they thickened the posts [2] and increased the strength of the beam (the architrave). Hence the vast solidity

---

[1] The order, I may say for the uninitiated, means the complete ordonnance of the column, the architrave resting immediately on its capital, the frieze and the cornice. It is the final expression of the simple device of the post and lintel, of the beam resting on the heads of two or more posts ; and there is little doubt that in its ultimate origin, the order is the translation into stone of the details of a rudimentary wooden construction.

[2] The columns of the Temple of Poseidon at Paestum measure in height a little over $4\frac{1}{4}$ times the lower diameter of the columns as against the 8 diameters to the height which became the normal proportion later. Choisy (*Hist. de l'architecture*, i. p. 313) says the columns at Paestum have no entasis. This

of the Doric order of the Temples of Sicily and Magna Graecia. The Greek was incurious about construction, *qua* construction. He found, in the column and the lintel, means perfectly adequate to realize his ideal of high unalterable beauty, and he was content. The Romans, who for a time were satisfied with these simple methods, became impatient of the constructive limitation of the post and lintel. They wanted to cover in great spaces, and to leave the floor unencumbered ; and concentrating on this they arrived at the arch, the vault, and the dome, and so became the greatest builders of the world. To them, the orders were a mere appanage of decoration, which they never properly appreciated, of which they mistook the intention, adopted the worst elements, and often enough made a gross misuse. The Greeks took another line. They adopted the column and lintel once for all as the only possible method of construction, and devoted all their labours to the incessant refinement of this type, eliminating the unessential, arriving by constant selection at the most perfect expression of their purpose, and their purpose was not that of the Roman and the modern architect, mainly utilitarian, it was directed entirely to the aesthetic appeal, the appeal to the emotions through beauty of line, of form, and in a less degree of colour. ' The whole fabric of Greek art goes to pieces when it is brought into contact with a purely utilitarian nation like Rome.'[1]

is wrong. The entasis in the columns of the temple of Poseidon is well marked. In the buildings called the ' Basilica ' and the temple of Demeter the diminution and entasis of the shafts are much overdone.

[1] *Hellenistic Sculpture*, by Guy Dickens, p. 85. The author,

Of the two orders, the Doric and the Ionic, the Doric seems to me the purest embodiment of the Greek spirit, in its faultless form, its austere restraint and rejection of the unessential. It was, moreover, the order *par excellence* of the Greek Temple of the mainland. The Erechtheion was the only Ionic temple of first-rate importance in Greece, and the employment of the Ionic order in Greece was confined to interiors and minor buildings. As for the Corinthian order, the favourite order of the Romans, it was scarcely recognized by the Greeks. In all their greater temples, in Greece, in Sicily, and Magna Graecia, they used the Doric order.

How this order was arrived at we do not really know. Ingenious conjectures have been made as to its origin in wooden construction, and though some of these conjectures are more probable than others they leave us pretty well where we were in regard to the stages by which it reached its final form. It has been suggested that the Doric column originated in the wooden post of the earliest temples, such as are supposed to have existed in the Heraion at Olympia. The square post would have its angles taken off, and become an octagon, and the further elimination of the angles would gradually bring it a form nearly circular in plan in which the arrises of the chamfered angles would remain, and this might easily suggest, to artists so sensitive as the Greeks, their further refinement and definition by a slight

who wrote with something of the insight of the artist as well as the accurate knowledge of the scholar, died of wounds, on the Somme, in 1916.

hollow between the arrises, which would constitute the flutings of the Doric column. Its derivations from the Minoan and Mycenaean columns seems most improbable. There are two essential parts in the Doric column, the shaft and the capital (the Greeks did not use any base for this order). The Minoan columns taper downwards instead of upwards, an utterly unconstructional form, and though in the Palace of Knossos columns of this shape may have been used to carry lintels, at Mycenae the column was not used for constructional purposes. On the other hand, columns of great massiveness tapering upwards had been used long before in Egypt; and, though there is evidence against it, it still seems probable that the suggestion of the shaft of the Doric column may have come from Egypt. The earliest direct intercourse of Greece with Egypt occurred in the reign of Psammetichus I (671–617 B.C.). The Greek colony of Naukratis on the west side of the Nile Delta was founded by Milesians about 650 B.C. and by the middle of the sixth century B.C. definite trade relations were established between Naukratis and the mainland of Greece. The Greek settlement at Daphnae on the eastern arm of the Nile appears to have been founded at about the same time as Naukratis, in both cases with the sanction and encouragement of the Egyptian king. The earliest Doric temples in Greece, Sicily, and Magna Graecia date from the end of the seventh century and early part of the sixth century. The nearness of date makes it probable that the shaft of the Doric order had its origin in the Egyptian column seen by some quickwitted Greek when trading in Egypt.[1] When we come to the capital of the column, the roles seem to be reversed, for we find nothing in Egyptian architecture to suggest the echinus moulding under the square abacus of the Doric column, whereas the Mycenaean column had a rudimentary capital which may have suggested the idea of the Doric capital, but the notable thing about it is that when we first come across the Doric capital, in Sicily and Greece, it is already far in advance of anything that had gone before it in Greece, and it is quite different from the columns of Egypt. In the Doric Temple of Corinth (650–600) the columns have already reached the type form, the tapered shaft with its entasis or slight convex curvature in outline, its massive solidity (the ratio is one of diameter to four and a quarter of height), and the bold parabolic curve of the echinus moulding under the abacus of its cap. In this form, the Doric column was an absolutely fresh note in architecture. Archaic though they were, these columns at Corinth show that the Greeks were already on the track of those refinements of form, those optical corrections and compensations, which differentiate Greek architecture from that of any other race. The exaggeration in the entasis of the Archaic column disappears, its tapering was diminished, its height increased, and the overhang of the capitals reduced, till in the Theseion (465 B.C.) and the Parthenon (450–438 B.C.) we reach the final inimitable type. The column, which at

[1] In the tomb of Beni Hassan there are octagonal columns supporting a square abacus which more nearly approach the idea of the Doric column than any of the suggested originals.

Paestum was not much over four times the height of its lower diameter, is now over five times, the great overhanging capitals are reduced to reasonable dimensions, the depth of the entablature is diminished, the axis of the column is slightly inclined inwards to give the impression of stability, the shafts have the slight curve or 'entasis' just sufficiently marked to prevent the outline of the column looking incurved ; the lines of the stylobate, or continuous base, on which the columns stand, and the entablature which they carry, have a slight rise toward the centre in order to correct the impression of the lines sinking in the middle ; the columns at the angles are thickened, because standing free with the light all round them they would otherwise appear smaller than the columns standing against the background of the building. Nothing was left to chance ; every aspect of the building, the relation of every part to the whole, and of the whole to its part, was studied profoundly, so that there should be no failure in its perfect harmony. Except in Egyptian architecture, and there to a much smaller extent, nothing like this had been done before. What the Greeks did was to formulate a rhythmical architecture, in which each part stood in a definite and considered relation to the whole, so that even in their ruined state these Doric temples give an irresistible impression of a great idea, a great architectural epic, in which each detail, however beautiful, was subordinated to the unity of the conception as a whole. It is this abstract quality which lifts Greek Doric so far above the ambitious art of later ages, and indeed

GREEK ARCHITECTURE 155

above all but the very finest work of any period of architecture.

Many attempts have been made to discover the secret of this wonderful perfection of proportion. That the Greeks had a system of their own, that they worked to definite ratios of dimensions and number, and employed graphic methods of determining their proportions, such as the use of triangles and the like to determine the limits of their designs, seems certain. But no contemporary account of any such system remains ; and all the explanations that are given are *ex post facto*, made by the theorists analysing existing buildings, not by architects designing new ones. Some four or five hundred years later Vitruvius compiled a treatise on architecture, in which, following the doctrines of the school of Alexandria, he expounded a Greek theory of proportion on the basis of the human figure. Vitruvius is obscure, and does not seem to have been certain himself whether the proportion of the parts of a design were to bear a relation to the whole, analogous to that of members of a human body to the body as a whole, or whether the proportions of the order were to be taken from the actual proportions of the human body ; and he complicates the position by reference to the ' perfect numbers ' of the Greeks. But here again he was uncertain whether the ' perfect number ' was ten or six. After which, and having in his reference to the human figure as the canon of proportion unwittingly set a trap for the scholars and artists of the Renaissance, he drops the subject, and digresses into a general classification of temples, with formal rules for the placing and dimensions of columns, which have formed the staple of treatises on classical architecture ever since. One should speak with gratitude of the labours of Vitruvius. After all, his is the only technical treatise left us on the subject; but he applied to the pure Greek temples a system evolved centuries later by critics and theorists; he was thinking chiefly of Roman versions of Greek architecture, and he was more interested in technical rules and precepts for the use of architects than in that abstract beauty which was all the Greek cared for. No classification, however laborious, will reach the mystery of Greek architecture. Its beauty is too subtle to be reduced to any formula.

The Doric order reigned supreme throughout the great period from the sixth till the end of the fifth century B.C. It failed with the failure of the high ideals of Athens. Other forces came into play to which it no longer responded, and later Greek critics even found fault with the Doric order for certain 'mendosae et inconvenientes symmetriae;'[1] but that order, the true symbol of the sons of Heracles, was one of the most momentous contributions ever made to the art of architecture. It was the keynote of Greek architecture through-

---

[1] *Vitruvius*, iii. 1. The difficulty was, that if the triglyph was placed on the angle of the building (the practice of the Greeks) and the next triglyph was placed over the axis of the column, the metope (or square panel) between these two triglyphs would be larger than the metopes between the triglyphs axial over the other columns. The Greeks solved the problem by reducing the width of the end inter-columniation, but later critics disliked this, and removed the end triglyph from the angle and placed it axial over the end column.

out its finest period. Later it was superseded by the Ionic order, and when Rome became paramount in the Western world that, in its turn, yielded its place of pride to the Corinthian order, opulent, luxurious, a little vulgar, a true register of the lowering of the sense and standard of beauty that followed the downfall of Athens.

Meanwhile, on the other side of the Aegean, the Ionic order was reaching its perfect form through a similar process of systematic thought on a type definitely adopted. The Greek colonies in Asia Minor were of very early origin. Legend attributed their foundation to the earlier inhabitants of Greece, driven out by the Dorians. By the sixth century B. C. the Greek colonies were well established on the east and south-east coasts of Asia Minor, and had evolved their own characteristic architectural idiom in the Ionic order and its column, more slender than the Doric, with its moulded base and its strange characteristic capital, unsuitable from the constructional point of view in stone or marble, yet ultimately attaining the exquisite beauty of line and modelling of the capitals of the Erechtheion at Athens. Two things seem fairly certain as to the origin of this capital; first, that it was derived from the wooden horizontal head-pieces fixed on posts to reduce the bearing of the primitive wooden lintels ; and secondly, that the first suggestion of the volute reached the Ionian Greeks from the East. A crude anticipation of the volute is found in Phoenician work, and it also appears on a Hittite relief at Boghaz Keni in the middle of Asia Minor. Its origin in either case was oriental, and we have here the other motive in Greek architecture, Eastern, at any rate exotic,[1] and, as compared with Doric, almost alien to the true Greek genius. Yet this astonishing people gave it a form as far removed from its barbarous originals as the Doric capitals of the Parthenon from the capitals of the columns of Mycenae, and when the Greeks of both sides of the Aegean drew together, after the defeat of the Persians, the Ionic order crossed the sea and assumed a place of honour in the temples of Greece, still, however, with rare exceptions, in subordination to the Doric order. In the Greek colonies in Asia Minor, the home of its origin, the supremacy of the Ionic order had long been recognized. The great Ionic temple of Hera at Samos, 368 feet long by 178 wide, is supposed to have been built at the end of the sixth or early part of the fifth century B.C., and this was the forerunner of the great fourth-century temples of Ionia built when architecture had changed its direction and the Hellenistic age was beginning its adventurous career.

With these two orders as the terms and idioms of expression the Greeks built up the architecture of their temples. Their plans were the simplest possible. The rudimentary type was a simple chamber or cella, with a loggia open to the air except for two columns standing between the two extremities of the side walls, which terminated in pilasters known as 'antae'.[2] The next stage was

[1] See *Cambridge Ancient History*, vol. ii, chap. xx, iii for the influence of the civilizations of peoples in the hinterland of Asia Minor on the Hellenic settlements on the Aegean coast.
[2] Vitruvius gives this as the 'aedes in antis'.

to bring the colonnade forward,[1] stage number three repeated the column at the other end of the building,[2] stage number four continued the colonnade along the sides,[3] stage number five doubled the colonnade on all four sides,[4] and stage number six retained the outer rows of columns but omitted the inner row along the sides, leaving a wide passage-way all round the main building.[5] Vitruvius gives a further classification by the spacing of columns which will be found in all the handbooks of classic architecture. With minor variations in detail, these types remained constant for the temples of Greece and Rome. The principal alterations occurred in the extension of the Temple proper, at the expense of the surrounding colonnade. In the archaic temples, such as the older temples of Selinus in Sicily (sixth century B.C.) the Portico and colonnade occupy three-quarters of the site. In the temple of Hephaestus (Theseion) at Athens (fifth century B.C.) the cella occupies a little more than half the total area, and in the Parthenon, built some twenty years later, the size of the cella is still further increased. Most of these temples were covered in. Hypaethral temples, in which the cella was open to the sky, are mentioned by Vitruvius, and it is probable that some of the larger ones at any rate were partly open to the sky. But how the openings were arranged is almost entirely a matter of conjecture.

[1] Pro-style (colonnade in front).
[2] Amphipro-style (colonnade at both ends).
[3] Peripteral (single colonnade all round).
[4] Dipteral (double colonnade all round).
[5] Pseudo dipteral (inner row of columns omitted).

The roof used was of a very flat pitch, 1′ of height to 4 of base, later it was even flatter, and this dictated the slope of the pediments. This roof covered the whole of the building, that is both the cella and the colonnades on either side of it, and as the Greeks were ignorant of the principle of the triangulated truss built up of beams in compression and tension, they were at a loss to know how to carry their roof without pushing out their walls.

Hence the great solidity of their buildings and such clumsy expedients as the central colonnade in the 'Basilica' at Paestum. In the temple of Poseidon the cella is only some 13 feet wide. One has to bear in mind, in thinking of Greek architecture, that the Greeks were not constructors in the sense that the Romans were; they built well and the best of their masonry was extraordinarily skilful: only by unusual skill in the cutting and setting of stone could they have carried out the delicate curves in the columns and other parts of their buildings; but construction, in the sense of the invention of new methods to meet difficult conditions, did not interest the Greek, and one cannot help thinking that the Greeks may have been more successful with the outside of their buildings than with the inside, and it seems clear that they devoted most of their attention to the external elevations. It is not really known for certain how they lit their temples, though of course all sorts of suggestions of top-lighting have been made.

In the temple of Poseidon at Paestum there is a clerestory order above the main order of the cella, the diameter of which seems to have been arrived at by continuing the diminution of the shaft of the lower order through the architraves, above which it emerges again as a very much smaller order. It is possible that in some cases they were lit only from the principal entrance, and it is certain that the Greek did not want for the interior of his temple any such floods of light as are necessary under our northern skies. In the first place, he enjoyed a brilliant and penetrating light, so that within his colonnades reflected light was amply sufficient to show the friezes and other ornaments, and he did not hesitate to use strong primary colours to heighten and explain their effect, wherever he found it necessary. In the second place, within the shrine itself, other considerations came into play. A certain luminous atmosphere, rather than positive light was what was aimed at, and the deep shadows of these internal colonnades might have helped this effect, adding to the mystery of the figure of the god.

This, too, may be the explanation of what must strike an architect as an anomaly of design, the Greek habit of placing enormous figures in the interior of their temples. The Greek, in his own way, was a very religious man. In his temple he was doing his utmost to set forth the majesty of his god, and if it was necessary for this purpose he was even prepared to sacrifice his principles as an artist, to ignore the scale of his interior and the rhythmic harmony of his design, by the introduction of gigantic figures. The eye judges by what it knows, and the readiest way of arriving at some idea of the size of a building or a monument is by relating it to the normal size of the human figure. Vitruvius, in his confused way, suggested that the human figure was the canon and standard of architectural design, but how is it possible to determine the scale of a building which contained a figure at least six times the size of a man reaching from the floor to the roof ? The chryselephantine figure of Zeus at Olympia, made by Pheidias, is supposed to have been some thirty-five feet high, and to have reached nearly to the roof, passing the double tier of columns and the gallery above the aisles of the cella. Moreover, the god was represented as seated on his throne, so that by no possibility could it have been in scale with the building so far as the architecture was concerned. Even the gigantic temple of Zeus at Agrigentum with its external columns 61 ft. 9 in. in height, and large enough for a man to stand within one of the flutings of the columns, could hardly have stood up to figures on such a scale as this. Such a violent contrast in scale broke the principle of συμμετρία, that strict relation of the part to the whole which the Greek artists maintained elsewhere with scrupulous care. Artists with such a consummate sense of proportion as the Greeks possessed would hardly have made a mistake here, and the conclusion one comes to is that where their religion was in question everything had to give way. Indeed, one can imagine the tremendous effect of this colossal figure seen dimly in the half-light of the cella, filling the whole temple with its presence. The same anomaly in scale occurred in the Akropolis at Athens, where the vast figure of Athene Promachos must have reduced the beautiful Caryatides of the Erechtheion to insignificance. M. Choisy makes a gallant effort to show that this want of relationship in scale, and also in the siting of the temples, was deliberate and considered. As a fact, the only rule that seems to have been observed in the time of Pericles was that new temples should always be built on the site of the older ones, but axis lines were neglected, and even the masses of the Propylea, beautiful building as it must have been, did not balance. The Akropolis was just a collection of unrelated buildings, and in the great Temenos of Delphi the various monuments were all anyhow.[1] The Sacred Way meandered about like an S, and the only method it observed was to clear the various treasuries and shrines which appear to have been scattered about within the enclosure, with a disregard of each other little less than brutal ; a rather suggestive symbol of the internecine rivalry of the small Greek states. At Delphi, also, there was a huge figure of Apollo Sitalkas, said to have been seventeen metres high,[1] which must have been hopelessly out of scale. The fact was that Greek architects of the fifth century had not yet arrived at the conception of the city as a whole. They had an admirable eye for a site, for example, the position of the Parthenon itself and the temple of Hera Lacinia at Agrigentum placed high above the sea, but it is unhistorical to invest even the architects of the Parthenon and the Propylea with a knowledge and outlook which was not thought of till a hundred years later. Even the Greek architects and sculptors of the fifth century B.C. were not omniscient, yet within their limits, in their mastery of what they set themselves to do, the artists of the age of Pericles remain un-

[1] See *Delphi*, by Dr. Frederick Poulsen, p. 52.

approachable, and theirs was the Golden Age of architecture. They had fixed for all time essential elements of the art, and had set up a standard of attainment in pure form which no subsequent architecture has ever been able to reach.

The fall of Athens closed this splendid chapter, but Greek architecture was by no means done with. The Silver Age, the Hellenistic art that followed, is of intense interest. With the rise of the Macedonian monarchy the stage of history shifted from the mainland to the Ionian colonies on the coast of Asia Minor. Cities such as Ephesus and Miletus became immensely prosperous; Mausolus of Halicarnassus, the Attalids of Pergamos, possessed wealth that would have been unimaginable to the Greeks of Marathon. The City State, fighting desperately for its existence, inspired by high ideals of patriotism and religion, was a thing of the past. These Greeks of Ionia were well content to enjoy the comfort and prosperity of a settled civilization without having to fight for it; and the whole atmosphere of their existence must have been different from the strenuous life of Greece in the fifth century. Moreover, the Ionian Greek, influenced, even if subconsciously, by the spirit of Asia, was by temperament unable to maintain the intellectual level of the Doric architecture of the mainland; and a difference appears in the whole orientation of art, in sculpture perhaps even more than in architecture. The history of Hellenistic art has yet to be written. It has been described as decadent, and it was undoubtedly responsible for some very poor stuff, but it also produced the 'Victory' of Samothrace, one of the finest things ever done in sculpture, and some very remarkable developments in architecture. It is not to be judged by the standards of the art that preceded it. The Ionian Greek of the fourth and third centuries B.C. broke away from the tradition of the mainland, a tradition always rather alien to his instincts. His interest lay less in a somewhat impersonal religion than in the assertion of his own individuality. He did not understand the lofty patriotism and the high ideal of abstract beauty that had inspired Pericles and his artists in the Akropolis. Indeed, there is a curiously modern feeling about much of his work, which became more marked as he came under the dominance of Rome. The individualism, the realism, the revivalism, and the commercialism of modern art were all anticipated by the Hellenistic artists of Ionia, of Rhodes, of Alexandria, and of Athens itself in the Roman period. Civilization was becoming more complex, and one finds this reflected in Hellenistic art, at once more florid than the Doric of the fourth century, yet also more skilful in its handling of complicated problems of planning and design. No one wanted archaic simplicity when the wealth of Asia was flowing into the treasuries of the Ionian States, and the expression of this opulent ease is found in their magnificent temples, such as the third temple of Artemis at Ephesus, of which the outer colonnade measured 342 ft. 6 in. by 163 ft. 9 in., or the temple of Apollo Didymaeus at Miletus, 165 feet wide by 360 feet long out to out of colonnades ; or the amazing monument of Mausolus of Caria at Halicarnassus, or the great altar of Pergamon. Fragments of the columns of the temple of Artemis, now in the British Museum, tell of its size and richness; they also give the first hint of the downfall of art and civilization which was to follow centuries later. The Greeks of the great period had kept the structural parts of their building free of ornament. It would never have occurred to them to interfere with the lines of the column in any way that would contradict its purpose ; but the Greek architects of Ephesus not only placed their columns on pedestals (making them so far less stable in appearance), but they adorned the lower part of their Ionic columns with figures, of admirable execution but perfectly inappropriate in the position they occupy. One cannot imagine Pheidias making a mistake such as this. Splendid in execution as Hellenistic sculpture often was—it won its place at the expense of architecture—one looks in vain for that power of selection and restraint which gives its undying distinction to the earlier work.

The Greeks of the fifth century realized that architecture is an art with a definite purpose other than that of a mere vehicle for sculpture, and that it makes its aesthetic appeal by its own inherent qualities of rhythm, and proportion, spacing, mass, and outline. Though they used sculpture and colour to heighten and intensify the effect of their architecture, they saw very clearly the function of the arts in relation to each other, and kept their sculpture and their colour in strict relation to the aesthetic purpose of their architecture. It is a point on which later architects went lamentably astray. A great deal of early Renaissance work is mere ornamentation of buildings, indeed in buildings such as the Certosa of Pavia the architecture has ceased to exist, and most of the bad architecture of the last fifty years is due to the deplorable fallacy that ornament is architecture. The columns of Ephesus, the sculpture of the altar of Pergamon, brilliant as they were in technical accomplishment, were the first hint of that decline which was in time to undermine the whole fabric of the arts. Architecture was deposed from its high intellectual dominance. It tended more and more to become a conventional affair, and it was an easy transition from the exuberance of Hellenistic art to the point-blank vulgarity of Roman ornamental architecture.

It was, however, inevitable that the fine simplicity of Periclean art should vanish with its ideals, and one finds a certain compensation in the extension of the range and outlook of architecture, which we owe to the Hellenistic architects of the fourth and succeeding centuries B.C. So far as perfection of form was concerned, it was impossible to carry the art beyond the stage to which Ictinus and Callicrates had brought it ; but there still remained something, and something very important, to be done. The Greeks of the fifth century seem to have had little conception of a city regarded as a whole composition of which the streets, the public buildings, and open spaces are the constituent elements. Axial planning, the consideration of the relation of building to building, seem to have been outside their consciousness, and each building was treated as an unrelated unit. But the inconvenience of this, its loss of opportunity, and the necessity of order and method, must have become apparent as civilization became more complex and more exacting. By the end of the fourth century B.C. the tradition of architectural technique was firmly established, and architects were able to turn their attention to problems of large planning, and these they seem to have handled with extraordinary skill. So far, what had been done in this direction had been due to religious inspiration, as in the processional ways leading to the Egyptian temples, or the avenue of figures at Branchidae. What the Hellenistic architects did was to think out consecutive schemes of city planning, in which the dominant motive of arrangement was artistic. They had learnt to treat the temples, the public buildings, the open spaces and approaches as the elements of one harmonious composition, in which the utmost use was made of the natural opportunities of the site. At Ephesus, for example, there is supposed to have existed a consecutive scheme, larger than anything of the kind carried out even in France in the eighteenth century, though the evidence, it should be noted, is largely conjectural. As presented by sanguine and enthusiastic restorers the scheme was magnificent. Next the port, and facing it on one side, was the Arsenal, a regular building opening on to a court surrounded by a colonnade, which again opened on to the great 'Place', a square enclosure some eight hundred and fifty feet wide north and south by six hundred and fifty feet east and west,[1] surrounded by a

---

[1] The Place Vendome measures 450 feet by 420; Grosvenor Square about 650 by 530, and Lincoln's Inn Fields about 800 by 630 measured from wall to wall of buildings.

colonnade on all four sides, with exhedrae, or semicircular recesses. In the centre of this Place was an oblong water-piece, about three hundred feet by two hundred, and on the farther side, opposite the Arsenal buildings, were the Senate House and other public buildings ; and behind these and to the right and left of them the Theatre and the Stadium, partly excavated in Mt. Coressus. The Arsenal, the great Place with its water-piece, and the public buildings were laid out on an axis line and on a regular rectangulated plan.

A scheme such as this (if it is possible to accept a conjectural restoration), thought out in all its bearings, meant a real advance in the range of architecture. It is useless to look for the faultless beauty of the fifth century, but the resourcefulness and skill of the Hellenistic architects give a new meaning to the art ; and indeed they might almost be said to have established the first stage in the development of its modern practice. It was from these very able Hellenistic architects that the Romans learnt the monumental planning of their cities, and for centuries the architects most frequently employed were Greeks of Asia Minor. At this point Hellenistic architecture merges into Roman and loses its distinctive character. Through Roman it passes on to modern architecture, and so in a sense the chain is complete ; but between this later art and pure Greek architecture there is a great gulf fixed, the difference is not only of technique, but of outlook, of ideal, and of temperament. The Doric of Paestum, Selinus, and Segesta, the Theseion and the Parthenon, remains for all time the perfect expression of the soul of ancient Greece.

It is one of the ironies of history that when, in the fifteenth and sixteenth centuries, scholars and artists awoke to the fact that there had been a great architecture in the past, they should have known of no other version of it but the Roman. What splendid developments might have followed if the finer spirits of the Renaissance, Alberti, Bramante, or Peruzzi, had founded their theories of architecture on the temples of Sicily and Magna Graecia instead of on the debased examples of Imperial Rome. They, at least, would have caught a glimpse of the beauty of abstract form and perfect harmony, the secret of which seems to have been revealed to the Greeks alone among the peoples of the world, and to them for only a transient period of their history. Unfortunately, when Greek architecture was discovered in the second half of the eighteenth century, it became the Shibboleth of the 'virtuosi'. The national traditions, both of France and England, were lost, Greek architecture became the fashion, and the misguided enthusiasm of pedants and amateurs insisted on a literal revivalism which completed the extinction of architecture as a vernacular art, and replaced it by the series of revivalisms from which it has suffered for the last one hundred and fifty years. Conscious and deliberate tinkering with the art of architecture has ended by destroying it.

We can never hope to revive Greek architecture, nor should we attempt to do so. There was once a well-known Scotch architect who held that the column and the lintel are the only permissible form of construction, and with this limitation and ill-selected Greek detail he produced some fantastically ugly buildings. Following a similar line of thought, a famous critic of the last century condemned methods of construction not sanctioned by the Old Testament. Both were wide of the mark, because above and beyond all technical details of architecture is the spirit in which it is approached, the intellectual outlook of the artist on his art, and this may express itself in widely differing forms. In Greek architecture of the Golden period that outlook was definite and distinctive, and it was one that has a very urgent lesson for us to-day. The aim and ideal of the Greek was beauty of form, and this beauty, which he sought in the first instance as the expression of his religion, ultimately became almost a religion in itself. To the realization of this ideal he devoted all his powers, sparing himself no pains in chastening his work till it had attained the utmost perfection possible. He merged himself in this work, without thought of the expression of himself in his vision of a divine and immutable beauty. It hardly occurred to him that his individual emotions were worth recording. (In the sculpture of the great period the expression of the face is one of unruffled calm.) Although religious emotion was the source and inspiration of his work, his work was impersonal. He was aloof from that feverish anxiety for self-revelation which has made much modern art so interesting pathologically, and so detestable otherwise. Nor again had he anything of the virtuoso about him. To him technique was never an end in itself. In Hellenistic art it became so, but not in the Golden Age. Indeed, he was sometimes almost careless of exact modelling, and in architecture he did not use the order as a mere exhibition of scholarship. In his search for beautiful form he stood upon the ancient ways, patient and serene, moving steadily to his appointed end. 'Ainsi procéde le génie Grec, moins soucieux du nouveau que du mieux il reporte, vers l'épuration des formes, l'activité que d'autres dépensent en innovations souvent stériles, jusqu' à ce qu'enfin il atteigne l'exquise mesure dans les efforts, et dans les expressions l'absolue justesse.'[1] There have been rare periods since, when architecture has moved with the same calm unhesitating purpose; Gothic architecture, for example, in the twelfth and thirteenth centuries, and certain phases of architecture in the eighteenth century in France and England, when tradition was still active and vital, and artists were content to let well alone.

Modern conditions seem to be wholly against the Greek standpoint in art. The arts are in the melting-pot, the old standards of attainment are trampled under foot, and the prophets prophesy falsely. Quite lately we were asked to find our inspiration in the fetishes of the Gold Coast, and if the aim of the artist is to outstrip his brethren in brutality, the advice is sound.[2] A recent critic

---

[1] Choisy, *History of Architecture*, i. 298.

[2] The latest folly is to deny that it has any value at all. We are told by one of those gentlemen with foreign names who enlighten our ignorance of art in the daily papers, that 'the fetish of Greek art is dead', 'killed by the patient labours of the

GREEK ARCHITECTURE

justified the antics of certain artists by the necessity they were under of advertising themselves. That, no doubt, is the readiest way to immediate success. But the question for the critic is, not the personal advancement of the artist but the value of his work ; and one would ask if any good work at any period in the history of art has been inspired by this ambition to shout louder than one's neighbours. Certainly the standpoint of the Greek was the exact opposite. He did not seek notoriety. He was happy with his inner vision of beauty, and intent only on its realization. He had not the smallest desire to shock or startle any one. There are occasions when shock tactics are necessary, but they are not necessary every day in the week, nor is it necessary to make a clean sweep of the past before one sets to work in one's own little corner of art.

What is wanted in modern art is some consciousness of this old Greek spirit, some recognition of its value. The Greeks of the age of Pericles wanted neither revivalism nor revolution ; they moved forward, without haste or anxiety, on traditional lines, and they were able to do so because their art was so interwoven with their life that, in the plastic arts, they could no more have changed their methods of expression than they could have changed their manner of speech. That high outlook on life is lost and hardly to be recovered under modern conditions of social life and political

critics '—the critics, one presumes, who also demand the demolition of Venice because it does not provide ' the ineffable joys of a sleepless life.' Noise and advertisement appear to be the ideals of the Futurist.

government. It was perhaps only possible under the true democracy of the small Greek city state, when every citizen took his share in the ordered life of the community. Yet the Greek ideal remains. In our fitful fever of honest intention and wrong judgement, high endeavour and point-blank commercialism, Greek art, the art of Pheidias and Ictinus, is still the wise mother to whom we must return. The lesson of the Parthenon is the lesson of a steadfast vision of beauty, held high above individual effort and failure, realizing itself not in complex detail or calculated eccentricity, but in serene and exquisite simplicity of form. It teaches us that in the arts there are no short cuts, and that anarchy, the destruction of what has been won for us in the past, is not advance, but the straight road to the bottomless pit of barbarism. Instead of repudiating the work of his fathers the Greek carried it on to its perfection, and built his palace of art on a sure foundation because he turned neither to the right hand nor to the left, but steadily set his face towards the light.

April 1921.

NOTE.—Mr. Wade-Gery (*Camb. Ancient History*, vol. ii, chap. xix, p. 525) says that the effect of the coming of the Dorians 'is almost wholly negative, they destroyed much and brought nothing'. His argument is based on patterns of pottery, swords, and brooches. My own impression is that inferences drawn from such evidence are by no means conclusive as to relative civilizations. Like conditions may produce like results without those results being necessarily related, and if these masterful invaders from the north did not create Doric architecture, I would ask, who did?

CHRISTOPHER WREN

ON February 25th, 1723, Christopher Wren passed away in his sleep ; ' cheerful in solitude, and as well pleased to die in the shade as in the light.' Except in the last few years of his life his career had been one of extraordinary distinction. Single-handed he had designed and carried through to completion the most splendid church in Christendom. In every kind of architecture as then practised the work that he did remains to this day our standard of attainment. Yet there is no evidence of his ever having received any specific training in architecture. He never went to Italy, and he was thirty when, if I may so put it, he jumped straight into architecture from the platform of the Savilian professorship at Oxford. How was this possible ? How did it affect his art ? It is a curious fact that, out of the six most famous architects of the seventeenth century, Inigo Jones, François Mansart, Bernini, Perrault, Jules Hardouin Mansart, and Wren, two, if not three, began as amateurs, without specialized training in architecture ; Bernini the sculptor who prided himself on translating architecture into terms of painting and sculpture, who designed that magnificent colonnade, and who very nearly let down the dome of St. Peter's ; Perrault the scholar, physician, and anatomist ; Wren the Fellow of All Souls and Savilian Professor of Astronomy. In 1662 Wren was generally recognized in England as a most remarkable young man, but his laurels had been won in mathematics and astronomy.

The men with whom he associated were men of science, not the specialized science of our day, but a science that formed part of the general culture of educated people. The inventive ability, which Wren had shown in his boyhood, had been exercised in ingenious mechanical appliances. One finds no reference of any kind to architecture, and it is doubtful if Wren had ever thought of it before, as a young Fellow of All Souls, he made the acquaintance of that intelligent, if rather priggish, amateur, John Evelyn.

It is true that Wren's position was exceptional, quite apart from the incalculable element of genius. Wren came of a good stock of the upper class. His father was a country rector, who afterwards became Dean of Windsor and Registrar of the Order of the Garter. His uncle was Bishop of Ely, and remained a prisoner in the Tower for eighteen years rather than abate his unswerving loyalty to the Throne. Wren himself was educated at Westminster ; at the age of thirteen he was able to write, in Latin verse, a dedication to his father of an astronomical invention ; at the age of fifteen he translated into Latin a treatise on dialling with the sonorous title of *Sciotericon Catholicum*; at the age of sixteen he wrote a treatise on Trigonometry, and while yet a boy was associating with Scarborough, Wilkins, Boyle, and Ward, the men who afterwards founded the Royal Society. In 1650 he entered, as a gentleman commoner, at Wadham College, Oxford, and four years later was elected a Fellow of All Souls. No wonder John Evelyn described him as ' that miracle of youth and prodigious young scholar, Mr. Christopher Wren '.

In 1657 he was appointed Gresham Professor of Astronomy in London, and in 1661 Savilian Professor of Astronomy at Oxford. In the *Parentalia* there is a list of forty-four tracts on scientific subjects written by Wren. By the age of thirty he was already recognized in England as one of the most distinguished mathematicians of his time and as a young man of most ingenious mechanical invention. So far there is no evidence that he had even thought about architecture ; but such was his reputation, and so great was the claim of his family on the royal favour, that when an examination and report on the fortifications of Tangier was required Wren was invited to undertake the work, and, though he declined the invitation, he was appointed Deputy Surveyor-General, with a promise of the reversion of the Surveyor-Generalship, to which he actually succeeded in a few months, on the death of Sir John Denham. He was instructed to examine and report on the work necessary in old St. Paul's and Windsor Castle, and to complete the Palace at Greenwich from the designs made by Inigo Jones some thirty years before.

The amazing thing is that Wren should have been entrusted with all this extremely responsible work though it must have been known to his friends that he was without any experience in architecture. In fact he owed the appointment not only to his brilliant reputation, but to the intrigues of influential friends, such as John Evelyn, working on the well-known inability of the English people to believe that more than one able man can exist at a time, and their habitual indifference to trained professional opinion, for it is a regrettable fact that Wren's introduction to architecture was the result of a discreditable job. The reversion of the post of Surveyor-General had been promised to John Webb, the pupil, assistant, and nephew of Inigo Jones, an experienced and able architect who also had claims to royal gratitude. Yet Webb was ignored, and the work handed over to a young man of thirty-five, who, however brilliant, could scarcely claim to be even an amateur in architecture. It was even worse than the somewhat similar intrigue by which Jules Hardouin Mansart superseded Le Pautre at Clagny in 1674. It almost takes one's breath away to find that in 1666 Wren, who so far had only designed Pembroke Chapel at Cambridge for his uncle, and the Sheldonian at Oxford, was appointed ' Surveyor-General and principal architect for rebuilding the whole city, the Cathedral Church of S. Paul's, all the parochial churches, in number fifty-one, . . . with other public structures.' (*Parentalia*).

However, rightly or wrongly, Wren had now got his opportunity, the opening which is even more necessary to the architect than to the painter and the sculptor, for the latter can leap into fame as it were out of space by some memorable work which he has been able to execute unaided, whereas the architect may have noble notions, but he has to wait for somebody else to find the resources to carry them out. All he can do is to labour patiently at his technique till his chance comes. But Wren, in this regard, was one of the favoured of fortune ; for, without previous training, a reputation won in other fields and influential backing placed him not merely in front of other architects of his time, but in a position of supremacy which was not to be disputed for nearly forty years. A position of such serious responsibility would have frightened a man of less confidence in his own consummate ability, but Wren's resource and incomparable quickness of intelligence enabled him to avoid disaster. He appears to have concentrated his study of architecture into something less than a year. A hasty wrestle with Serlio's very inaccurate *Architettura*, and Fréart's still more inaccurate *Parallel of Architecture*, recently translated by his friend John Evelyn, appears to have concluded his technical studies, followed in 1665, the year of the plague, by a six months' stay in Paris, just at the time when Colbert was urging Louis XIV to complete the Louvre, when Le Vau was already superseded, and when Bernini, brought against his will from Italy with almost royal honours, was vainly endeavouring to carry his design for the Louvre in face of the determined and organized opposition of the French architects. Wren had excellent introductions. He met Bernini, but, as Wren describes it, 'The old reserved Italian gave me but a few minutes' view. It was five little designs on paper; for which he has received as many thousand pistoles'. Wren mentions as the leading architects of his time the elder Mansart, Le Vau (whom he calls Vaux), Le Pautre, and a certain Gobert, who is now forgotten; in other words he only met the older generation and none of the younger men whom Colbert was to bring to the front in the service of Louis XIV. However, Wren did his best. He visited Fontainebleau,

St. Germains, 'the incomparable villas', as he calls them, of Vaux, Maisons, Ruel, Courans, Chilly, Effoane (? Ecouen), Rinçy, Chantilly, Verneuil, and Liencour, great houses familiar in the views of Perelle, and Versailles, not the Versailles of J. H. Mansart, but the 'petit château de Cartes' of Louis XIII, 'not an inch within but is crowded with little curiosities of ornament'. There is no suggestion in the *Parentalia* of that close personal research such as De L'Orme and Inigo Jones undertook in Italy. Instead of this Wren brought back 'all France on paper'; in other words he had recourse to the pattern book, the equivalent in his time to that fatal short cut to architecture to-day, the interminable photograph. 'All France on paper' must, in 1665, have consisted of the engraved work of the elder Marot, views by Israel Sylvestre, and those sets of designs, those 'Lambris à l'Italienne', 'Cheminées à la Romaine', 'à la Moderne', 'à peu de frais' and the like, which year after year from 1654 onwards Jean Le Pautre produced with indefatigable regularity. Hardly the training required for a great architect, scarcely enough to qualify him as an amateur. Contrast this with the training of any of the famous French architects of the time of Louis XIV, men who, without a tithe of Wren's genius, were familiar with every corner of their technique. Contrast it with the training of Wren's predecessor, Inigo Jones. Although little or nothing is known of the early life of Jones, it is certain that his training in the arts began early. 'Being naturally inclined in my younger years to study the arts of design, I passed into foreign parts to converse with the great masters thereof.' Unlike Wren, Inigo Jones came from a tradesman's family. We know nothing of his education, and the tradition is that in early youth he was apprenticed to a joiner. It is not known how he was able to make his first visit to Italy. Perhaps, like the first students of the French Academy in Rome, he made his way through France on foot. But his great natural ability, his skill as a draughtsman, and his profound study of his art in Italy, speedily made his reputation. On his return to England he revolutionized stage scenery, and became known to the Court as an artist of great learning and resource. In 1613–14 he spent another year in Italy, devoting himself almost entirely to the study of architecture as handled by the Italian writers, more particularly Palladio, and as illustrated by actual buildings in Italy. In 1615 he was appointed Surveyor-General, being then forty-two years of age, and two years later he designed the Queen's House at Greenwich, in its own quiet way still one of the best examples of Neo-Classic in England. Wren's equivalent building would be the Sheldonian Theatre at Oxford, designed in 1662 though not actually built till 1668. These two buildings illustrate the whole difference between the work of the trained artist and the work of the able amateur, and in saying this I make no criticism of the incomparable genius of Wren in his maturity. In the Garden Front of the Queen's House Inigo Jones made no attempt at parade or ingenious detail. The façade is of two stories only, rusticated up to the first floor level, plain from the string-course up to the entablature. It is surmounted by a simple balustrade; there are no figures, not even finials or urns. In the centre on the first-floor level is an open Loggia, in five bays, with Ionic columns. With the exception of the capitals of these columns there is not a scrap of carving on the building, or any attempt at ornament beyond the balustrades and simple mouldings, yet such is its admirable proportion, its fine play of light and shade, and its calculated reticence, that this façade is a perfect little masterpiece. Only years of study and a thorough mastery of his technique could enable an architect to stay his hand at the perfect point of balance.

The Sheldonian Theatre, completed when Wren was thirty-seven years of age, is about as bad as it can be. The interior is commonplace and chiefly remarkable for a well-constructed queen-post roof. On the outside the building is just a great lump. The outer walls enclose the theatre without any attempt by the designer to build up a monumental composition, and the detail is exceedingly bad. The rusticated arches are too high for their width, and Wren appears to have forgotten his own plinth in calculating the height to the springing. The upper story fails in being too high for an attic and too low for a story proper and the mouldings are coarse and ignorant. As for the south front with its slender Corinthian order below and its atrocious composite pilasters above, the misfits of its members, and the evident anxiety to make the façade imposing by swags, cartouches, ' Dolopumque exercitus omnis ', it is, I suppose, the worst piece of architecture perpetrated in Oxford before the days of the Gothic revival in the last century. The grotesque terminal figures on the front to the Broad, if by Wren, were a reminiscence of what he had seen at Vaux, but they had no relation to the building or anything else. Indeed it is quite evident that at this date, say 1668, Wren had still pretty well everything to learn in regard to the technique of architecture.

Similar solecisms of design appear in the earliest actual building of Wren, the Chapel of Pembroke College, Cambridge (1662). The Corinthian pilasters in the front are some 12 diameters high. If Wren had even consulted his Serlio he would have found out his mistake, yet Wren made his pilasters 12 diameters, I am convinced, from simple inexperience, not as an experiment which in his latter years he would have justified. In the turmoil of the Civil War the fine achievement of Inigo Jones had been forgotten and English architecture might have developed on different and most interesting lines, had it not been for the violent arrest of the lead given by that great architect. My impression is that, in these early days, apart from construction, Wren did little more than give general indications of architectural details, relying on the traditional skill of English masons trained in the school of Inigo Jones. No Jacobean carver, for instance, could have executed the capitals of Pembroke Chapel, or the bold swags to the cartouche of the pediment, and the debt of Wren to his great predecessor in these matters both of design and craftsmanship has never been fully recognized. Wren's genius sometimes blinds us to the fact that he was never quite certain of his technique ; for example, the unsatisfactory treatment of the angles of the octagon under the dome of St. Paul's, and of the colonnade at the west front, where the bays at either end have nothing to do and bear no relation to the organic structure of the design.

Yet technique is not the last word in architecture ; behind it all is what a man has to say. Has he any vision of his own, any fresh outlook on facts, any new version of their relations, to contribute, and again, what sort of personality does his work reveal to us ? For an architect can express himself in his work quite as much as the painter and sculpture in theirs, otherwise the office boy would be as good as his master. The architect's likes and dislikes, his aims and ideals, his temperament and outlook on life, are written on his work in ineffaceable letters. The mean meticulous man will be mean and meticulous in his architecture; the great man will be great and the small man small. There have indeed been phases of architecture in which the architect seems to have set himself to say as little as possible in the most correct language, for example the Frenchmen in the time of the French Revolution, and some of our own men a little later. This is better than the ignorance and lack of invention which characterizes much of our modern architecture, but it is empty, unconvincing stuff, a mere simulacrum of the vitality of great design. Fine technique is essential, but one wants something more than this; and one wants technique, not as an end in itself, but as a means to an end, the adequate realization of ideas, that perfect fusion of thought and expression that was once shown within its restricted limits in the Parthenon.

Yet here, again, we must recollect that some problems of architecture are much more complicated than others ; that in the Parthenon, for instance, perfect as it is as far as it goes, there were no problems of planning and construction comparable with those handled with consummate skill by the Romans in their Thermae, and by Wren in St. Paul's. In modern architecture something more is wanted than a perfect sense of form ; resourcefulness, imagination, quickness in appreciating the opportunities of the problem ; in a word, invention. It is on this ground that I believe Wren to have been, on the whole, the greatest architect known to history. The faults of technique in his earlier work are easily seen, and he never possessed the perfect mastery of form of Inigo Jones, François Mansart, or the younger Gabriel, but no architect has ever approached him in the range of his imagination and his invincible resource. No one has ever assimilated an acquired manner so completely as Wren, making of it a true vernacular and national art. The faults due to the lack of regular training were corrected by his experience of actual building. That which no amount of regular training could have given him was the quickwitted intelligence with which he grasped each fresh problem, the fine temperament which expressed itself in his design, so that it is no dead thing or vain repetition, but cheerful and serene, instinct with life, and convincing of its reality.

Wren may have picked up some details of ornament in his brief visit to France, but there was no one in France who could have taught him his plan for the rebuilding of London. In the *Parentalia* there is a description of Wren's scheme of 1666, which shows that, with one curious exception, Wren anticipated all the devices of the modern town-planner ; the vista, axial planning, through communications, the organized relationship of street, square, and building which ought to govern the laying out of a city. Wren's principal streets were to be 90 feet wide (Regent Street is only 80 feet), others were to be 60 feet wide and lanes about 30 feet, 'excluding all narrow dark alleys without thoroughfares'. The widest street in Henri IV's scheme for the 'Porte et place de France' in Paris was 36 feet, and as late as 1758 when Carpentier prepared his great scheme for Rouen the principal street was to be only 36 feet wide.

The same splendid grasp of opportunity is shown in Wren's completion of Greenwich, as compared, let us say, with that monument of accomplished mediocrity, Versailles, to which Jules Hardouin Mansart kept adding block after block till all semblance of fine composition was lost, and the Palace justified St. Simon's verdict, ' la main d'œuvre est exquise en toutes genres, l'ordonnance nulle'. So again in churches, the French had lost the art of the smaller Parish Church. They were skilful in what a French writer calls 'the implacable façade' of the Jesuit Church, but their design nearly always followed the same lines, or if not it dropped into the vulgarity of the Paroisse at Versailles. Wren, who had to design in haste under most difficult conditions of site and money, produced, in the towers and steeples of his City churches, a series of buildings unique in the grace of their design, and their happy adaptation of means to ends. He followed no particular precedent, but just invented as he went on, and one of the most striking features of his designs is their individuality, the audacity with which he broke with tradition in some places, and recalled it in others. It was perhaps fortunate for Wren that he never went to Italy, and that his stay in France was so short, in the sense that he trusted to his own invention. The only architecture that he was really familiar with was the architecture of his own country, and there is a marked Jacobean feeling in the designs of some of his towers and steeples ; it is indeed this combination of severer architecture with the caprice and fancy of an earlier manner that gives to much of his work its peculiar charm.

I have pointed out the scanty training that Wren received in architecture and the technical shortcomings which were the result of it in his earlier work. Yet he advanced *per saltum*, and one of the most astonishing things about him is the rapidity with which he picked up knowledge on the scaffolding of his buildings, learnt from his own mistakes, and developed his natural power of mechanical invention into a faculty of architectural design almost without a parallel in range and resourcefulness. When Inigo Jones began the restoration of old St. Paul's, he started with a noble Roman Portico, but he never reached the problem of the rest of the building. François Mansart, past master as he was, did one beautiful little church in the Faubourg St. Antoine, and much admirable domestic work, but he worked under exceptionally easy conditions, his supposed nephew Jules Hardouin had behind him all the wealth and resources of the most powerful monarch in Europe, and the finest craftsmen of the world to carry out his designs; but his vast palace is architecturally inferior to Hampton Court, and there is no real comparison between his Church of the Invalides and St. Paul's. Wren, unaided, and with little technical training, had to deal with problems of design and construction which these men never attempted, and he dealt with them with astonishing success, pressing into their service not only his science and mechanical skill but the broad sagacious outlook of the humanist, so that in all his mature work there is a certain balanced equipoise, and a happy freedom from the preciosity which some better-trained architects have not always been able to avoid.

His advance from the warrant design for St. Paul's to the design actually carried out is one of the most astonishing things in the development of any artist. The first design proposed by Wren was a fine idea, though not very practical for the purposes of service. Jules Hardouin Mansart borrowed it in his design for the Church of the Invalides, but it would not have done for St. Paul's. Wren's design for the elevation was immature and not very satisfactory, and the design was rejected through the influence of the clergy. It appears from the *Parentalia* that Wren proposed various alternative designs, and in the result the authorities, finding themselves wholly ignorant of the subject, appear to have tossed up as to which design they should select. The words of the Royal Warrant (1675) are significant : ' Whereas among divers designs which have been presented to us we have particularly *pitched upon* one, as well because we find it very artificiall, proper and usefull, as because it was so ordered that it might be built and finished by parts '. It was in fact one of the most preposterous designs ever made for a cathedral, with its ill-designed portico, its commonplace flanking towers terminating in candelabra, and the enormous dome cut short half-way up and continuing in a drum surmounted by another smaller dome running up into a telescope steeple. I have never quite understood how Wren could have seriously put forward this design, except on the assumption that when he first attacked this immense problem he was technically unequal to it and had only a glimmering of its ultimate solution. The main ideas of St. Paul's were vaguely present in the design, the central dome and the great west portico with its flanking towers, but their expression was crude beyond belief, and it was only Wren's quickness in advancing from point to point in design, his freedom from the obsession of fixed ideas, and his readiness to learn on every hand, that saved him from the fiasco of the warrant design, and finally reached the glorious masterpiece of St. Paul's as we now have it. It is too often forgotten that architecture is not only a difficult, but also a very laborious art, which requires a longer and more persistent apprenticeship than any of the arts. There is no short cut to it, no such thing as the brilliant improvisation, no escaping the incessant critical analysis of one's own work. Wren was great enough to realize his own defects, and throughout his career he steadily used his ample opportunities to correct them. His case, if ever there was one, bears witness to the soundness of the ' gymnastic ' view of education, the view, that is, that the business of education is not to take a boy young and cram into his half-developed mind quantities of specialized facts, but to keep on training his mind on broader lines, so that at a later stage he is able to attack his specialized technical studies with the disciplined intelligence of the trained soldier. To make a boy specialize at sixteen is like throwing a child into deep water before it has learnt to swim. Few people possess the natural endowments of Christopher Wren, but even Wren could not have mastered his art with the rapidity that he did, had he not previously received a thorough education on the broadest possible lines.

To us to-day Wren is a figure of the greatest significance in his development as an architect, his actual work, and his personality. To the serious architect, to the man who regards architecture, not only as a means of livelihood, but as a very great art, there are two problems yet unsolved— the training of architects, and the manner of expression of modern architecture. In spite of repeated efforts for several generations we are still not quite happy about the training of our students, and we seem to be still uncertain how we are to express ourselves. In our methods of training we veer wildly from concentration on detail to its entire neglect, from the minutiae of revivalist Gothic to the impassive futility of Neo-Classic, and we set our students face to face with these problems before they have attained to the knowledge of good and evil. Where we are wrong is, I believe, in closing down general education too soon, and in forcing on specialized training. It is impossible ever to catch up this initial defect. Wren began very late, but he possessed an admirably equipped intellectual machine and simply raced through stages that would have cost years of struggle to minds less thoroughly trained and disciplined.

It will be said at once that Wren was a man of genius and therefore exceptional, but this only points to another mistake that is made in assuming that anybody of ordinary intelligence can become an architect if sufficiently trained. The extremely scanty results of our vast apparatus of training, state-aided or other, prove the contrary. It is forgotten that an architect, that is the designer, as distinguished from the business man and the building policeman, is first and last an artist in building materials ; and a still more costly mistake has been made in forgetting that an artist, if he is worth his salt, is as rare and exceptional a person as a poet. Wren was one of these rare and exceptional persons and this was the other factor in his extraordinary career.

Wren's work is so familiar that it needs no

panegyric, but I would suggest that he was a true modernist in the best sense, that he met each problem squarely on the merits of the case ; and did not attempt to twist it and torment it to suit a formula. The men who succeeded him, Campbell and the Burlington clique, affected to despise his work because it did not conform to the practice of the ancients, just as 100 years later the architects of the Directorate built a temple of Mars if they had to design an arsenal, and a temple of Aesculapius if they were asked for a school of medicine. To Wren such pedantry would have seemed childish and ridiculous. If he erred at all it was in the opposite direction. In all his work he showed the strong practical sense and freedom from affectation which has always been the best tradition of Englishmen, and which perhaps they alone possess among modern nations.

In his skill as an architect, his lofty outlook, and his freedom from the sins of egotism and ambition that do so easily beset us, Wren remains the nearest approach I know of to the ideal architect. The technical details, on which he is open to criticism, are insignificant in comparison with his greatness both as an architect and as a man. His large conception of his art, his extraordinary ability in plan and construction, were far above the level of the merely learned technician, and behind it all was Wren, the man himself, the gentleman and the scholar, wise, humorous, and equable, a delightful companion, an artist aloof from self-seeking and advertisement. One does not think of Wren as one does of the younger Mansart, thrusting his way to the front, elbowing rivals out of his path, betraying his friends, ending with a marquisate and a salary of twenty to thirty thousand pounds a year, and dying only just in time to escape complete disgrace. All Wren received was a salary of £300 a year as Surveyor-General, and part of this was withheld from him. He ended his days in retirement, almost disgrace, thrown out of office by the intrigues of knaves and charlatans, but in his lifetime he enjoyed the affection and esteem of all the best men of his time, and he left behind him the finest reputation ever enjoyed by any artist of this country.

January 1923.

Wren's City Churches

Since the date when this paper was written, the measure known as the Union of Benefices and Disposal of Churches (Metropolis) Measure, 1924, has been passed by the Church Assembly almost without opposition. The effect of this measure is, among other things, to establish machinery by means of which City churches may be demolished, the sites sold, and the money so realized applied to ecclesiastical purposes elsewhere. The measure has yet to receive legislative sanction. If it does, and unless it is materially altered, the future existence of any of these churches is no longer safe. 'Against a determined attack, they might go down like so many ninepins'.[1] The Commission of 1919, appointed by the Bishop of London, with Lord Phillimore as Chairman, dealt with forty-seven churches, and

[1] Leading article in *The Times*, Nov. 21, 1924.

recommended the removal of the following nineteen :

All Hallows, Lombard Street (Wren, 1694).
All Hallows, London Wall (the younger Dance, 1764-9).
St. Botolph, Aldgate (except the Tower) (the elder Dance, 1744).
St. Katherine Coleman (Horne, 1739).
St. Clement Eastcheap (Wren, 1686).
St. Dunstan in the East (except Tower) (Wren 1698 ; Laing, 1817).
St. Magnus the Martyr (except Tower) (Wren, 1676 and 1705).
St. Mary at Hill (Wren, 1672).
St. Mary Woolnoth (Hawksmoor, 1716).
St. Michael Cornhill (except Tower) (Wren, 1672, and Tower, 1722).
St. Alban, Wood Street (Inigo Jones, about 1635 ; Wren, 1685).
St. Anne and St. Agnes (Wren, 1680).
St. Botolph, Aldersgate (1790).
St. Dunstan in the West (except Tower) (J. Shaw, 1830-2).
St. Mary Aldermanbury (Wren, 1677).
St. Michael Royal (except Tower) (Wren, 1694 and 1713).
St. Nicholas Cole Abbey (Wren, 1677).
St. Stephen, Coleman Street (Wren, 1676).
St. Vedast (except Tower) (Wren, 1694-7).

The Commission appears to have made up its mind beforehand as to what it meant to do, no matter what arguments were advanced to the contrary, for though the incumbents were asked whether they wished for an oral interview, that interview was only granted ' when this wish and our view that the particular church might be removed *coincided*' (Report, April 26, 1919). Nor, with the exception of Lord Hugh Cecil and Sir W. J. Collins, do they appear to have had any misgivings, for they state (clause 16) that they had ' gone carefully into the question of the architectural merits ', &c., of the several churches and had ' come to the conclusion that those named in the list might *well be removed* '.

With the exception of six, all these churches are by Wren. Of the rest, St. Mary Woolnoth (Hawksmoor) is a very fine design, unique in its way in this country. All Hallows, London Wall, is a remarkable example of the younger Dance, who designed old Newgate, and St. Botolph, Aldgate, with its excellent tower and steeple, the best thing ever done by that robust old person, his father, the architect who designed the Mansion House. St. Magnus the Martyr, St. Nicholas Cole Abbey, and St. Vedast, are characteristic examples of Wren at his best, and it must not be forgotten that, between 1781 and 1897 seventeen churches, all of which were designed by Wren, had already been removed. A Commission appointed by the Bishop of London in 1899 recommended the removal of ten churches, but only two of them so far have been destroyed. It seems that the pace was not fast enough for the authorities. Lord Phillimore's Commission conceived that they should proceed ' on different and bolder lines ' and accordingly recommended the destruction of these nineteen churches, Lord

Hugh Cecil and Sir W. G. Collins dissenting. The report of the L.C.C. made in 1920 says, ' If the recommendation of the Bishop's Commission to destroy nineteen more churches be adopted, the result will be almost to double the number of churches removed during the last two and a half centuries '. The Commission apparently went on the principle that where two or three churches are gathered together one at least should be removed. They left towers in certain cases, regardless of the fact that the tower and its church are one composition, and that to pull down the church and leave the tower is nearly as fatuous as the suggestion that these churches might be taken down and re-erected elsewhere under different surroundings. The Commission seemed to be unaware that old buildings by long use and association acquire a personality, as it were a soul of their own, and that they cannot be torn up by the roots and transplanted bodily. The Commission estimated that the demolition of the nineteen churches would set free for Church purposes an income from benefices of £23,931, and produce a capital sum of £1,695,600 from the sale of sites, and in addition would save an annual sum of £1,461 for repairs, and £4,858 for maintenance of services. These weighty financial considerations appear to have impressed the Bishop and his advisers so favourably that in the Report of the Metropolitan Churches Committee issued in December 1922 (the Bishop of London Chairman), no comment was made on this proposal of wholesale destruction, and it is only the strong and sustained protest of those who care for these churches that has brought the matter into the open.

There is unfortunately no doubt that the Church is badly in need of money for many excellent purposes, but is this the right way to find it ? The loss of these beautiful buildings, so admirable in design and so rich in associations as many of them are, is not to be measured in terms of money. But apart from this aspect of the matter there are serious objections in principle to the measure as it stands.

1. In the first place its promoters assume the right of the Ecclesiastical Authorities to regard these churches not as a trust for posterity, but as property to be disposed of when and as they think fit. Now most of the churches in question were not built out of ' the ancient resources of the Church ' at all, to use Lord Hugh Cecil's phrase. The greater part of the cost of the fabric was defrayed by the Coal-tax and most of the beautiful fittings and furniture was provided by parishioners. Moreover, these churches are maintained not by tithes but by rating.

2. In the second place, if a legalized machinery for the demolition of churches is established, one does not see where it is to stop. The Bishop might ultimately think it his duty to annex St. Paul's Cathedral. As was said by *The Times*, ' if this measure is passed hardly any of the City churches will be safe '.

3. No mention is made by the promoters of this measure of the fact that it simply wipes out the existing power of veto possessed by patrons and vestries under the Union of Benefices Act of

1860. The report of the Metropolitan Churches Committee, dated December 1922, says, 'we substitute in the Measure the scrutiny and control of a large Board' and believes that 'the control will be exercised with a wider outlook and with a more wisely proportioned judgement than the old veto of the vestries and patrons'. But why should the Bishop assume this? The only evidence so far as to the probable action of the Benefices Board is the report of Lord Phillimore's Commission in 1919 recommending the destruction of nineteen churches, and the report of the influential Committee of 1922, from which I have just quoted, which tacitly acquiesced in this destruction.

> If they do these things in a green tree
> What shall be done in the dry?

Surely the opponents of this measure are justified in their fears for the churches. Moreover, 'substitute' seems scarcely the right word for what is in fact the high-handed suppression of a legally constituted right.

4. It is assumed that the City churches do not pull their spiritual weight, yet I have heard the Rector of St. Michael's, Cornhill, describe with much eloquence and in detail the admirable work done by the City churches in aiding the spiritual welfare of the many thousands of people who have to work in the City. With one or two exceptions these churches are havens of rest to many a hard-driven worker, and if these could only bear witness to what the churches mean for them there would be less talk of demolition. The fault is not with the churches, but in the use that is

made of them. I know that the Rector of St. Michael's is a strong advocate of a greatly extended use of the churches, and in all that concerns the appropriation of any of these buildings to spiritual purposes other than that of the ordinary service I think the Ecclesiastical authorities would have the sympathy of the strongest opponents of this measure. What one feels bound to resist is the legalized destruction of the churches.

5. We may dismiss at once the crude suggestion that these churches can be taken down and re-erected elsewhere stone for stone. It is not the physical difficulty that is in question, but the spiritual. Buildings belong to the site on which they stand and cannot be transplanted in this way.

6. It appears that a principal object of this measure is to find money for the building of churches elsewhere—that is, the old church is to to be destroyed and with the proceeds of sale a brand-new church (probably Neo-Gothic) is to be built in some district where the people are too indifferent to raise the money themselves. This is not how the old churches were built, and might be thought to reflect on the spiritual efficacy of the Established Church itself. Why not begin with some modest shelter and work by degrees, as was suggested in 1919 by Lord Hugh Cecil himself, instead of attempting these ambitious schemes? It is the disastrous principle of the 'dole' over again and a dangerous precedent that may ultimately recoil on the Church itself. The measure has been very properly described as 'unimaginative'. It is one of those desperate

short cuts which seem nowadays to be the accepted remedy when any difficulty occurs.

The opponents of this measure, who represented all those societies and laymen who care for the beauty and associations of our historic monuments, asked in the first instance that a power of veto should be given to an independent and representative body of lay opinion, either the Ancient Monuments Board or the recently established Royal Fine Art Commission. It was pointed out that this did not mean obstructionism, and it was conceded that cases might occur in which the church in question was architecturally and historically so unimportant that reasonable men might feel bound to consider the question of its destruction, but it was urged strongly that even then its destruction should only be permitted after the fullest consideration had been given to the case in all its bearings, not only by Ecclesiastical authorities but by the best expert lay opinion. The veto asked for was not granted, but a reference to the Royal Fine Art Commission in the early stages of the inquiry was provided in the measure. As, however, it was also provided that the report of the Fine Art Commission was to be submitted to the Benefices Board, and the latter were empowered to turn it down then and there if they thought fit, the concession was of doubtful value, and left us still uncertain whether expert lay opinion is to receive any adequate consideration. Lord Hugh Cecil invites us to believe that 'a predominant part in the deliberations of the Board shall be in the hands of wise, moderate, and broad-minded men who will carefully weigh all artistic and archaeological considerations'.

Yet Lord Phillimore's Commission deliberately reported that nineteen out of forty-seven City churches might 'well be removed'. The Benefices Board, the Bishop's Committee, the Bishop himself, might be fair-minded men fully conscious of their responsibilities all round. On the other hand they might not, and as the only evidence so far available is the report of the Phillimore Commission of 1919 and that of the Bishop's Committee of 1922 we are compelled to think that they might not. *The Times*[1] was justified in saying 'this measure goes to Parliament frankly as a means for destroying churches and not for preserving them and everything in it is heavily weighted in favour of clerical opinion'. Even the apparently harmless Union of Benefices may mask an attack on the City churches. As has been pointed out by one of the City clergy, 'If several churches are grouped round one control church, their fate is a mere matter of time'. The closer this measure is considered in all its bearings the more dangerous it appears to be. Lord Hugh Cecil in a very able letter to *The Times* (December 9, 1924) urged that the precautions provided in the measure to ensure the fullest inquiry are so many and so effectual that we may rest assured as to the safety of the churches. It may be so, but on the other hand it may not, and the evidence of the past shows that the effect of this measure is likely to be very different from what Lord Hugh would persuade us it will be. The issues at stake are too serious and far-reaching to admit of a gamble on an off-chance.

[1] *The Times*, Nov. 21, 1924.

ARCHITECTURE AND DECORATION [1]

I HAVE been asked to address you on architecture and decoration, but I am not quite sure that I am the right man to say anything on this subject, because, in defence of architecture, I have sometimes found it necessary to maintain that it is an art complete in itself ; that is to say not dependent for its impact on the emotions on any art external to itself ; and that its highest appeal is made through the abstract qualities of rhythm and proportion, mass and silhouette, and what the French call ' ordonnance '. But here at any rate I must sing my palinode. The history of art shows that however content an architect may be with the completeness of his own art, mankind in general wants something more. When the cave man had found his cave he decorated its surfaces with admirable drawings of the animals that he hunted ; all the stages by which man has worked his way out of barbarism into civilization are marked by advances in the decorative arts, and we must take it as an axiom that, whether we want it or not, man will decorate his buildings somehow. The problem for the artist is not only to draw well and paint or model or carve, but to use his brains in co-ordinating this with architecture, so that the two or more arts all pull together towards one preordained end. The Greek Doric Temple was probably the most perfect example that has ever

[1] An address delivered before the Royal Academy, January 1923.

existed of architecture and decoration, because the architect, the sculptor, and the painter worked with complete understanding of each other's limitations and resources.

Architecture, of course, touches every form of permanent decoration, the sculptor, the painter, the worker in metal, in stone, in marble, in wood, in glass, in mosaic, almost every art and craft ; but I am going to limit my few remarks to the sculptor, the glass painter, and the painter proper. And first, of sculpture.

The first rule that comes to my mind is one of Aldrich, the famous Dean of Christ Church, who wrote a manual of civil architecture about 130 years ago. Aldrich's advice in regard to sculpture was, not to have too much of it. ' Caelatura nimia venustatem opprimit ', ' too much ornament crowds out beauty ' ; and this was what happened in the earlier Renaissance, when the sense of architecture was lost under the embroideries of the ornamentalist, and was not recovered till really competent architects took charge, men such as Bramante, Peruzzi, and San Michele in Italy, Inigo Jones in England, François Mansart in France. Those interminable arabesques, panels, and pilasters, covered with ornament very skilfully executed but destitute of meaning in relation to the organic structure of the building façades, such as the Certosa at Pavia, or the Madonna dei Miracoli at Brescia, are not architecture at all ; and few things have done more to retard the art than this misconception of ornament as architecture. Sculpture of this kind is the expression, not of an artist's mind, but of a tradesman's skill paid by the yard. Indeed one knows of examples in which the ornament ends abruptly, presumably with the pay ; and I recollect a François Premier doorway at Bourges garnished by the carver with words which purported to be Latin, but were in fact gibberish, though doubtless they were intended and accepted by the owner as evidence of his Latinity.

If sculpture is used at all on a building, it should be used with a definite meaning. If you look back on the history of architecture, you will find that the great periods of the art are precisely those in which sculpture has been most under restraint—Egypt and Assyria for example, with their direct religious purpose ; the Greek Temples of the finest period, in which the sculpture was confined to the pediments and the metopes of the frieze and did not invade structural members ; thirteenth-century Gothic in a less degree, and the architecture of the middle of the eighteenth century in France and England, the work for example of Chambers or the younger Gabriel. It is in the weaker and immature periods that sculpture got out of hand, breaking all measure, degenerating into vulgarity and even nonsense, as one may see in the conventional trade ornament of modern buildings.

The Church of St. Vincenzo in Rome, by Martino Longhi, shows the disastrous effect of overloading. The sculpture is skilful enough. The architectural detail is learned, yet the effect of all this accumulation is almost barbaric. With a single piece of well-placed sculpture and one pair of columns and one pediment instead of three, the effect would have been infinitely better.[1]

This matter of the placing of ornament is of the first importance. Good ornament in the wrong place is worse than no ornament at all ; and the architect is directly responsible for this. It is no use placing sculpture where it cannot be seen, or is seen under conditions which reverse its values, or where the space is so contracted that the sculptor has no room to let himself go. For example, the sculptor will not thank the architect if he finds the recess provided for his figure is too small, or the pedestal impossibly bad, as happened at Perpignan the other day. A monument was to be erected to commemorate a local worthy; the design was entrusted to an architect and the bust to a sculptor. The monument was completed, and the date of the unveiling arranged ; but when the sculptor saw the pedestal provided for his bust he was so disgusted that he departed with the bust, declined to hand it over, and the local authority are now going to sue the sculptor. Again, though it rests with the architect to allocate the work, he ought not to impose tasks on the sculptor which are wrong in principle, such as terminal figures clearly unequal to the work, or sculpture which stultifies the original intention of the architecture. The columns of the Third Temple of Artemis at Ephesus are a familiar example of fine sculpture in

[1] SS. Vincenzo and Anastasio, 1600. This church has three Corinthian columns on each side of the central window, the two inner columns slightly advanced in front of the one behind with three breaks in the entablature, and three pediments one inside the other. It is exceedingly clever and quite absurd. See Ricci's *Baroque Architecture*, pl. 7.

the wrong place. The lower part of these columns is surrounded by figures in relief, streaming round the columns, and weakening the impression of enduring strength, which is obviously the very essence of a column. Pheidias would not have thought of such a thing, and Ictinus would not have allowed it. The twisted columns of Bernini's Baldachino, in St. Peter's, are another familiar instance of misapplied skill. In this matter of the determination of ornament the architect and the sculptor ought to pull together from the first, and having come to a working agreement they have next to consider the purpose of the building and the all-important matter of scale. It is obvious that the purpose of a building, whether e. g. it is to be a church or a gymnasium, must determine the general character of its decoration, and ordinary common sense can deal with this, but the question of scale is far more difficult, and indeed it is in this matter of scale, more than anything else, that we are apt to go wrong. However great an artist's natural endowment, the sense of scale has to be painfully acquired by much study of buildings, and meditation on sculpture in relation to buildings ; and no amount of designing on paper will build up this sense in the architect, nor will modelling in the studio, however accomplished, take its place for the sculptor. It is here that we are all so seriously handicapped by the want of tradition. Down to the end of the eighteenth century artists were saved from glaring faults by a well-recognized standard of technique and by accepted principals of design. All this was lost in the general upset of tradition which finally resulted in the rise of the Romantic school, and the artistic anarchy that has prevailed ever since. Moreover the training of architects, two or three generations back, had much to answer for. As a consequence of the unfortunate misconception of architecture generally known as the Gothic Revival, our attention used to be concentrated on detail, mouldings, capitals, and carvings. Nowadays the training in our schools seems to have gone to the opposite extreme, and to have become academic in the wrong sense. It seems to me that sculptors made a similar mistake in isolating their work, and limiting their technique to modelling in their studios. Even in the days when a tradition of monumental architecture did exist, serious blunders of scale have been made, and it has always seemed to me that one of the reasons why St. Peter's looks smaller than it is is that the figures in the spandrels and pendentives are much too big.

In monumental architecture the whole is greater than the part; it is far more important that the whole composition should hang together than that any part of it should be elaborated in the vain hope of saving a loose design by some one masterly detail.

Owing to a variety of reasons, which I cannot here indicate, there has grown up, in the last hundred years, an unfortunate division in the arts. The painter paints his easel picture, the sculptor models his figure or his bust, and the architect designs his buildings, each of them working in his own corner without thought or care of the other. We have among us excellent sculptors, and, may I say it, not incompetent architects, but in our schools of art, and elsewhere, the two arts are out of touch and they ought to be brought into much closer relationship than has been the case for many years. It was a fatal thing for sculpture when it left the builder's scaffold and retired into the studio, and when the architects forgot the heroic precedent of Greece. And in saying this I am not thinking of medieval sculpture with all its rare beauty, because we at any rate can never get back to the emotional and intellectual standpoint of the Middle Ages : I have in my mind the example of two famous men, Bernini and Alfred Stevens. Bernini is much in the fashion at the present day, and both as a sculptor and as an architect was a man of undoubted genius and profound knowledge, to whom modern sculpture, through French art, is indebted for some of its best, as well as some of its worst, qualities ; its best, in its freedom and revolt against pedantry, its worst in its failure in selection and constant tendency to bathos on the one hand and excessive emphasis on the other. Bernini prided himself on imparting the quality of painting to sculpture. In music, one has seen the disastrous result of this mixing up of the arts, and Bernini, who set himself to dissolve architecture into sculpture and to present both in terms of painting, was a pioneer on a very dangerous path. But his work cannot be overlooked. I will only mention two characteristic examples. In the monument of Pope Urban VIII in the Tribune of St. Peter's at Rome, the motive is melodramatic, almost vulgar, the ornament is overdone and the female figures are sentimental and silly, yet the monument as a whole makes an imposing composition, full of vitality and instinct with a splendid swagger, if I may use the term, which at any rate saves it from insipidity.[1] In the high altar of St. Maria del Popolo it appears to me the design would gain greatly by the omission of all the figures above the pediment. Bernini could never stay his hand. Though very skilful in all his work, there is a constant search for sensationalism, an effort to impress the spectator with the amazing ability of the artist, rather than to appeal to his emotions. All really great art is impersonal, in that the artist is far more concerned with the work that he is doing than with what may be thought of himself. But that it is possible to combine architecture and sculpture with complete freedom, yet without sacrifice of the essential qualities of either of those arts, is shown by the work of Alfred Stevens. Other men may have been his equals in the technique of modelling, a very few may have possessed his immense knowledge of detail, but I doubt if any sculptor, since the greatest masters of the Renaissance, has approached him in the architectonic sense, in the power of thinking in terms of mass composition, and of sustaining his work at that high level of concentrated thought. It is not enough to be able to model an exquisite figure, if the setting is feeble and the architecture wrong, and if one comes away from the monument with the impression of some one beautiful detail, and the rest nowhere, the monument has surely failed of its purpose. The Stevens Drawings Nos. 217,

[1] See Ricci, op. cit., pls. 84 and 108.

223, and 231, now being exhibited here—designs of walls with doorways between Corinthian pilasters—are the most masterly architectural drawings I have ever seen, and I would suggest them as models to our younger architectural draughtsmen who seem to be more intent on producing attractive water colours than on the study and presentation of architecture. This great artist conceived of his work as a whole, and not as an aggregation of details ; and it is in this more than in anything else that we architects and sculptors have still so much to learn.

One more point before we leave sculpture. There is a fashion now prevalent of presenting sculpture in blocks and squares. This work appears to be inspired by the art of primitive peoples, who got as far as they could with their posts and blocks, but were necessarily headed off by want of knowledge, want of the proper tools, and not least of all by the limitation of their own vision, and the harsh conditions of their environment. In those to us unimaginable days Terror was the dominant idea, and the hideous and the grotesque the only possible presentation. To attempt to translate modern thought and ideas into terms of expression which were almost inevitable in the twilight of civilization appears to me to be the merest affectation; and there is one permanent objection to these raids on primitive art, that the results have no sort of relation to the interiors of our buildings and convert them into museums ; a figure can have little decorative value if it is hopelessly out of harmony with its setting. The proper setting of the South Sea

Island god would be on its own altar surrounded by the relics of the last cannibal feast. To introduce him into our modern buildings is like letting loose a ' salvidge ' man in a drawing-room ; for, however much we may gibe at our civilization, it is impossible to divest ourselves of the traditions and associations on which it is built.

Stained glass occupies a curious, and in some ways anomalous, position in modern decorative art. In days when there was no printing-press, when the majority of people could neither read nor write, the glorious imagery of the windows of those great cathedrals, Chartres or Bourges, of churches such as Fairford, were as the very book of life, revelations by artists as convinced of their literal truth as were the people who worshipped below them. It was a genuine and sincere form of art. But however earnest the modern glass designer may be, it is almost impossible for him to maintain this level of complete conviction ; and whereas to the man of the thirteenth century St. Christopher in the window was a fact and a reality, to the man of the twentieth he is a piece of symbolism, representing an idea, a version more or less archaeologically correct of a beautiful legend. Instead of the implicit faith on which the medieval artist could rely, the modern artist may find sympathy in certain mystical temperaments, but in all the details of his representation he has to reckon not with the implicit and child-like faith of a people as convinced as himself, but with a critical and even pedantic archaeological opinion. The result is that ecclesiastical decoration is apt to stand in the same relation to decorative art as 'the hymns that we all sing', to use Matthew Arnold's unkind phrase, stand to poetry. The fact is that certain phases of art belong to special periods and special states of mind, and it may be that the art of the glass painter is one of them. There is another difficulty that he has to contend with. Unconsciously, he and the wall painter are in direct antagonism, in that the one deprives the other of the light that is necessary for the effect of his work. It is very little use finding spaces for the painter, if, as soon as he has finished his work, the light on which he reckoned is reduced by fifty per cent. ; and it would seem that those who have the custody of our churches and cathedrals should make up their minds whether they want their decoration to go on their walls, or in their windows. If they elect for painted decoration on the walls they should at least allow the painter light enough to show his work. If they elect for windows, they should limit the work on the walls to mosaics which, by the nature of their material and the necessary insistence on line in their drawing, are able to tell their story in the half light, half shadow of a dim-lit church. Much of the impressive effect of the mosaics of St. Mark's, or St. Vitale, is due not only to their severely abstract design but also the mystery of their forms, half suggested, half lost again in the shadow. I am assuming, of course, that the building in question is not a classical building, that it is one set out to catch all the happy chances of romance, not one designed on a complete and consecutive scheme in which the masses and planes, the voids and solids, the wall spaces, and the areas of light and shade—in short a classical design—are thought out as such from the first. I do not think in these latter there is much opportunity for the use of painted glass, and it is better away. The plain white walls and the clear glass of St. Paul's are still a regretted memory with some of us, a setting in which every spot of accidental colour, vestments, uniforms, the gleam of brass against the background of oak, told with magnificent effect. Yet some very beautiful modern stained glass work has been done by English designers and craftsmen, far finer, in my opinion, and far more suitable to its purpose than any that I have seen on the Continent or elsewhere. That there is and will be a demand for it I have no doubt ; all one asks of our artists in this regard is that in designing their windows they should think of them as elements in one great harmony.

Lastly, we come to the Decorative Painter : and here I speak with fear and trembling, in the presence of painters, not that it matters what I say because Mr. Clausen, in the next lecture, will put it right. Yet an architect must be allowed some voice in this matter, because after all it is his building, the creation of his brain, that is being decorated, and without his building there would be no painting at all. The architect cannot be relegated to the position of the boy who blows the bellows. There have been, it is true, artists who have made it their business to create the illusion that there was really no building at all ; the perspective makers of the end of the seventeenth century did quite a large business in painting awkward pieces of wall to look as if they were vistas of magnificent architecture, and I have seen an illustration of a remarkable ceiling by Pozzi in which it is hard to say where the building leaves off and the ceiling begins,[1] and I have stood in churches in Rome and been quite unable to decide whether certain architectural features in the vaulting are real or painted. You will find many of these compositions in the architectural engravings of men such as Jean le Pautre, or Daniel Marot ; and the fashion lasted well into the eighteenth century. Indeed, I recollect seeing somewhere a horrible French wallpaper, said to have come from the Exhibition of 1851, in which this manner still survived. As a matter of fact, architects cannot help feeling some little anxiety on this point, because painted mural decoration, when it once gets beyond the limits of a repeating pattern, may involve the illusion that the wall is not there at all ; the wall becomes a subordinate, almost negligible, affair in relation to the story told by the painter. In such interiors, for instance, as the Arena Chapel at Padua, this position seems to have been accepted. All the architect or builder was called in to do was to provide sufficient expanse of wall for Giotto and his pupils to tell the whole history of the Virgin and Christ. The interest of such work is immense, but it hardly comes within the scope of architectural decoration, and it seems to me to cut the problem rather than solve it. The interior of a building ought not to be considered as one gigantic

---

[1] See St. Ignazio at Rome by Pozzi, illustrated in Ricci's *Baroque Architecture*, pl. 69, and St. Matteo at Pisa by the two Milani (*c.* 1720, Ricci, pl. 68). The centre panel of the vaulting of the nave of the Jesu is another example.

picture; the two arts of painting and architecture are not really united when one of them has to be sacrificed entirely, and the better way is for the architect to endeavour to realize from the first the general problem of his scheme, the relation of plain to decorated surfaces, and to reserve in his design the spaces which he wishes to have painted, such, for example, as panels, lunettes, domed surfaces, and the like, definitely marking these off from the constructional features of the building. Within the spaces so allocated he would leave the painter a free hand, and make no further protests on the score of illusionism. One of the most beautiful examples I know is Benozzo Gozzoli's decoration of the chapel of the Riccardi Palace at Florence.

A point, however, does arise, in regard to which it is essential that the architect and painter should agree. Apart from the difference of treatment necessitated by the difference of purpose (as, for example, that of a church and that of a town hall), different modes of architecture carry with them their correlatives in painting. For example, those four glorious panels of Tintoretto, in the Ducal Palace at Venice, require for their setting all the audacious ornament that surrounds them. One can hardly imagine them in a setting of Greek Doric without some suspicion of discord. If the painter wants an orgy of colour he must have an orgy of architecture with it; scrolls, cartouches, twirligigs, volutes, lots of gilding, sumptuous ornament, mirrors, and marbles. This is what Lebrun and his associates did at Versailles, and one comes away from their work with an impression of much magnificence, if also a singular absence of ideas. Contrast this with Mantegna's exquisite work in the Camera degli Sposi at Mantua, where the architecture is of the simplest, yet perfectly wedded to the decoration. You may recollect the Putti supporting a tablet over the door, and the central panel in the ceiling opening to the sky, the first and one of the most beautiful examples of the open sky motive in the decoration of ceilings. Imagine this austere and noble art transported, let us say, to the Foyer of the Opera House in Paris, or lost in the corridors of Versailles, and one realizes that, in great decoration, the architect and the painter must go hand in hand. In a more florid, yet excellent, manner there is a remarkable example of painted architecture and figure subjects by Marco da Sienna on the walls of the Sala Paolina of the Castle of St. Angelo at Rome, and of course there is the tremendous Ceiling of the Sistine Chapel.

There is one other matter in which the painter might with advantage work more closely with the architect, and that is, in the design of the architecture that he introduces into his painting. It is sad to find imaginative visions spoilt by quite ignorant architectural drawing, columns of impossible proportions and shape, entablatures all wrong, arches that would not stand up, and so on. The great Italian decorative painters always took the trouble to master these details thoroughly, as, for example, Veronese in his *Family of Darius* and his *Marriage of Cana* or Tiepolo in his splendid *Antony and Cleopatra* in the Palazzo Labbia at Venice. To Tiepolo, most delightful of decorative artists, nothing was impossible, but I think the masterly architecture of the Labbia Fresco was directed by his friend, Mongozzi Colonna, architect and sculptor. Architects would readily place their technical knowledge at the disposal of their colleagues, and I would suggest to students who contemplate decorative paintings that they should study the external forms of architecture more closely. They might at least be put through their facings in the Orders of architecture, if for nothing else for the reason that this would introduce them to methods of working to some definite system of proportion. I believe it to be quite as important for architect and painter to work together in decorative schemes as it is for architect and sculptor, and I do not think our colleagues need be afraid of us. After all we are all out for the same thing, the quest of beauty, and when as architects we have asked our colleagues to consider with us these points of scale, tone, and character, we will wish them God-speed in their voyage to the Fortunate Isles.

We do not all live in palaces, and municipal buildings are not the only ones that cry out for decoration ; and for those few old-fashioned people who prefer to live in a house of their own there is still the opportunity of decorating their houses with something more interesting and individual than the latest thing in wall-papers. You will notice in the present exhibition some admirable examples of wall decoration on a less ambitious scale, designed for small interiors and the decoration of houses. I believe that a future is opening out here for our painters, the signs of which are full of promise. Down to the end of the sixteenth century wall-paintings on quite a modest scale were usual in our domestic buildings. There is a good example in a little room in the Flushing Inn at Rye, one side of which is decorated with camels, elephants with castles, and other delightful beasts. This art disappeared in the time of Charles I, when representatives of great collectors, such as the Earl of Arundel and the King himself, ransacked Italy for works of art. The easel picture superseded the wall painting. The double-cube room at Wilton, probably the finest room in England, was the greatest achievement of this time, so exquisite in its walls, the combined work of two great masters, Van Dyck and Inigo Jones ; so lamentably failing in its ceiling with its oblivion of the scale of the room. After the Civil War an attempt was made by Verrio Laguerre and Thornhill to reproduce the glories of the Grand Manner as practised in France ; and then the easel picture took charge again, sweeping away the feeble efforts of poor Angelica Kauffman, and the more vigorous work of such men as Ricci, examples of whose art you may see in the staircase and in the fine ceiling of the Assembly Room of Burlington House. The Grand Tour completed the business, and since then the dealers have arisen in their might and turned most of the great houses in this country into picture galleries and museums, with much profit to themselves and the result of the practical extermination of the decorative artist. It is fair to say that a minor contributory cause was the introduction of hand-painted Chinese wall-papers in the eighteenth century out of which very soon grew the highly profitable industry of the printed wall-paper which has held the field ever since, and has formed a useful background to our prints and pictures. But there are signs that decorative art is waking up again, and that out of all the welter of experiment of the last twenty-five years there is emerging a definite tendency in one direction, a fresh ideal of the painter's work. At one time it was held to be his business to produce mere illusions of nature. That idea has gone, and the reaction to the opposite extreme is going too, the ridiculous view that it is the painter's business to paint non-representative forms, that is forms totally divorced from any resemblance to nature; for example, that if a head is round or oval, you should paint it as an oblong or square. That view also is done with, and what as it appears to me the best of our younger men are doing is to avail themselves to the full of all the skill they can command in draughtsmanship, colour, and composition, and then deliberately to select, in order to give to their idea its clearest and simplest expression. If as a layman I may say so, I think they are setting their course in the right direction, and speaking as an architect I am sure that this is what is wanted for the decoration of our walls. We do not want realistic illusion, we do not want the effects of the kaleidoscope—what we do want is an art as sane and as disciplined as architecture itself. There are examples of such art in this exhibition, both figure work, eminently suitable and desirable for the small house as well as for the great, and work less ambitious than that, but in its own way not less delightful. People have sometimes sneered at 'Chinoiseries' and 'Singeries' as foolish and meaningless. It all depends on what you want in your wall decoration. If you insist upon it that your wall decoration must announce some great moral, intellectual, or religious truth, of course the Singerie is impudent and ridiculous; but one cannot always live at the level of the highbrow. When I am tired and worried I do not wrestle with the lucubrations of Signor Croce, but retire to the romances of Stevenson, or Mr. Buchan, or the memoirs of some *franc vaurien* of the eighteenth century. So it is with one's surroundings. If one wants something pleasant and companionable, easy to live with, interesting enough when one looks at it, yet not insisting on being looked at, the Singerie and the Chinoiserie have their modest use, with their humour and caprice and their dainty play of line and colour. There is an example in the Burlington Hotel in Burlington Gardens, not of the highest quality it is true, yet charming in its way, and if you examine the finer French examples, the work of Huet, for example, or Berain, you will find that they are based on subtle and accomplished drawing, and that within their unambitious limits they are genuine artistic work. So too are such decorations as are to be seen on the sides and soffits of the window openings of the galleries of the Doria Pamphili Palace at Rome—delicate arabesques in soft blues and yellows and reds on a grey-white ground with vignettes interspersed—and one should not overlook the best work of the Pompeian decorators. I am not for a minute suggesting that these examples should be copied. I am endeavouring to indicate lines of possible development. The condition of course of such work is that there should be no question of technical ability, the artist must be in no difficulties with his drawing and his painting, for it is of the essence of his work that it should be fresh and spontaneous, the one unpardonable offence is stodginess and incompetent technique.

I seem to have wandered a long way from my starting-point, for after introducing my remarks with a condition that they are made without prejudice to the independence of architecture, I find that I am suggesting, as one motive for its decoration, a manner chiefly distinguished by the charm of its flippancy and irrelevance, but the position is not so illogical as it might seem. The Palace of Art has many chambers, and my contention throughout has been that each of its chambers has its own appropriate treatment, waiting to be discovered by the artist, and that the right way to discover it is for the artists to come out of their hiding-holes, and to search for it together.

At one time I was so fascinated by Colbert's magnificent achievement in the reign of Louis XIV that I thought the State might organize this work, but we have learnt our lesson in regard to State interference, and a very costly one it has been, and I am convinced that there are no short cuts, that if the arts are to recover their lost positions it must be won back, not by State organization, not even by the 'patient labours of the critics', but by the steady purpose of artists themselves. Perhaps architects have more opportunities than other artists of breaking through the cordon of the middle men, and I would appeal to them to lose no chance of going direct to the individual artist and craftsman. It is only in this way that we can escape the insincerity of that 'art by the mile', which makes some fastidious minds recoil from all idea of decorative art.

January 1923.

OFF THE TRACK [1]

'SOME THOUGHTS ON ART'

I AM much honoured in having been invited to address the Birmingham and Midland Institute this evening, and if I cannot help feeling some misgivings at having to face so large an assembly I am encouraged by memories of an occasion long past, when this great city offered its hospitality to the last of the Congresses of the National Association for the Advancement of Art. It was in the autumn of 1890, thirty-four years ago, that a band of enthusiasts descended on Birmingham, intent on making art a living thing, and its training a reality, on breaking down the strongholds of prejudice and conventionality, and on bringing the Arts into touch with the actual facts of daily life. I recollect the place was full to overflowing, one member billeted himself in his sleeping-bag—John Brett lectured on Art, Onslow Ford on the Education of Sculptors, Graham Jackson on Architecture. The proceedings wound up with an eloquent address from Sir William Richmond, in which he appealed to Birmingham to dedicate one small building ' To the service of beauty ' in which ' the architect should be the master-builder, the sculptor the mason, and the painter the decorator '. The only immediate result of that lofty appeal was a unanimous resolution of the Congress that the Govern-

[1] An Address delivered before the Birmingham and Midland Institute, June 1924.

ment Art schools were ' inefficient as a means of training in individual art ', that they ought to be reorganized, and that closer relation ' should be established between artists and manufacturers—laudable objects, so far imperfectly realized. Yet this generation owes more than it knows to the enthusiasts of the 'eighties and 'nineties. Those were the days in which the Arts and Crafts Society and the Art Workers' Guild were founded by a body of young artists impatient of the conventions they found in the art around them, not quite so wise perhaps as we thought ourselves, yet aiming high, and William Morris, so splendid in many ways, so elusive and disappointing in others, was our leader. The high enthusiasm of those early days faded away, as it was bound to do, yet I believe that future historians of the art of this country will find that it was a real movement, the beginning of a break-away from the insipidity of the art of the third quarter of the last century, a movement which has not yet gone very far—but which is, I believe, through much affliction and with many aberrations, slowly moving forward. In a remarkable preface to the Arts and Crafts Essays in 1893 Morris wrote, ' the lack of beauty, which in the earlier part of the century was unnoticed, is now recognized by a part of the public as an evil to be remedied if possible '. The Pre-Raphaelite movement, with which Morris was closely connected, was itself a heroic protest, if not necessarily in the right direction. At least the state of placid satisfaction of the 'seventies, the sense that everything was for the best in the best of all possible worlds, was rudely broken up. The energy of the younger generation initiated an era of obstinate questioning and experiment, the results of which are still uncertain and obscure, but which were, at least, the first steps towards clearing away the dry-rot of Middle Victorian art.

Since those days much water has flowed under the bridge. The New English Art Club, now relatively so staid and respectable, began its turbulent career, and looking back some thirty years one gladly pays one's tribute to the pioneers of that movement, Charles Furse, Wilson Steer, Walter Sickert, and other reformers who perhaps hardly foresaw the full extent to which they were swinging the pendulum. Then there was the famous Rodin dinner, followed by the revolt against academic art, and the organized attack on the Royal Academy, conducted with much zeal but with a regrettable lack of candour. And within the last few years there has been the reorganization of the Academy itself from within, inspired by a definite anxiety to do justice to all sincere artistic effort, and to free the Academy from the reproach which attaches to all Academies, of being the *injusta noverca* rather than the mother of the Arts. All these episodes are of interest, and have an important bearing on the present position, but they belong to what I must call the politics of Art, rather than to Art itself. What I am now concerned with is the position of Art as an element of life. As an artist, one has to inquire whether the view of its function sedulously advocated by latter-day critics is right in theory, and whether, in point of fact, the hierophants of Art are not seriously off the track.

The subject is difficult and obscure, and one can only feel one's way to some sort of working hypothesis, and endeavour to find some sure ground amid all these conflicting views. The public has reason to be bewildered when it turns its attention to the Arts. The critics say one thing, artists say and do another, and though what the critics say leaves the artist cold, it is an unfortunate fact that the written word prevails with a public which, in matters of which it has little knowledge, is apt to place its faith in the daily paper, especially when it is faced with a total discrepancy of practice. Mr. Jones, the famous R.A., paints what he sees ; but Mr. Smith, the no less eminent exponent of non-representative Art, paints what he does not see ; and when the plain man goes to a gallery and finds the last word of non-representative art to be like nothing at all, he not unnaturally washes his hands of the whole business, and wraps himself in a cloak of stolid indifference to all the Arts. More sensitive people cannot do this. They are left with an uneasy feeling that something is wrong, that it cannot be true that modern Art means turning one's back on the work of the past, and declining to recognize its priceless legacy. Are we to forget Titian and Tintoretto and find salvation in Gaugouin and Matisse ? Are we to place on equivalent planes, being as we are children of the West, the sculpture of the great Greeks, Italians, and Frenchmen on the one hand, and the lumps which imitate the stocks and stones of archaic, and even barbarous, art on the other ? Are we to feel towards modern German architecture as one does to the architecture of Ictinus or Peruzzi, François Mansart or Wren? Surely not—yet modern critics seem to think we should. Indeed, the more foolish among them announce that the only thing to do with the mighty men of old is to forget that they ever existed. It is one of the unfortunate results of the loss of tradition that we have now no clear ruling as to the purpose of Art, and no definite standard of attainment recognized by all. If we had, I fear we might be tempted to make a clean sweep of a great deal of modern work in all the Arts, not so much because of any great failure in ability as of a total lack of direction. It is not fair to set the failure wholly down to the artist. Modern conditions are against him, not only because of the dissipation of interests and the restless impatience which results from the pace of modern life, but because the Arts no longer make the one clear appeal that they did in the past, in the days, for instance, when the fresco, the carved image, and the painted window were for the great mass of the people the only key to the world of the imagination; or as in those later days when, if the artist made his appeal to the few, those few were people of intelligence and power, who cared for the Arts and used their position to shelter and encourage them. We suffer in all the Arts from the incessant multiplication of strange gods. No sooner is one altar established than another is set up against it. It is the very diversity of appeal of the Arts nowadays that is one cause of their weakness; but the real danger of their disintegration arises from the persistent misrepresentation of the meaning and purpose of Art, and it is this that we have to bring out into the open and do our best to end or mend.

Artists are apt to put the blame on the public. I do not think this fair. Our national architecture shows that our people in the past were not insensitive to Art, and I believe myself that there are many more people in this country who actually care for beauty than we are sometimes allowed to suppose. We used to be told that, as a people, we are unmusical, which is clearly absurd to any one in the habit of attending concerts of chamber or orchestral music. I have noticed since the war that if music of any approach to excellence is being played in the streets of London by ex-bandsmen and other wandering minstrels, people of all sorts will stop and listen. I believe it is the same with the graphic and plastic arts. The public—and by that I mean the ordinary intelligent man or woman—does appreciate fine simple work when they are given the chance, but we have to admit that fine work is rare, and that much of the work advertised with such persistence and expounded with such ingenuity by critics is not Art at all, but at its best a more or less ingenious intellectual exercise, a well-meant but technically incompetent effort, at its worst what I shall say of it later. The real Philistines among us are those who call on us to bow down to gods whose feet are made of clay, on pain of excommunication if we fail to fall down and worship.

I am thinking of quite recent happenings and will give you two instances. The fashionable artist just now is Vincent Van Gogh. This is how a critic in a leading weekly paper writes of him.

He is describing the painting of a rush chair in a rather squalid room. ' Painted by Van Gogh,' he says, ' this rush chair becomes an object of profound significance, eloquent of the life of those who have used it, of the clumsy hands that have shaped it, and above all of its essential nature as a chair.' This is as bad as Ruskin at his worst. Taking the passage in order, the chair was a bad one for its purpose, the shaping was of the clumsiest ; and as to ' its profound significance eloquent of the life of those who used it ', this would apply just as well to a pint pot or an old boot. What is mischievous in this exposition is the introduction of irrelevant matter, the appeal to sentimentalism, in order to give a fictitious value to a somewhat unskilful and not very interesting study of still life. The portrait of ' the Postman ' in the same exhibition would surely have convinced any candid person that Van Gogh simply could not do what he wanted to, because he did not possess the necessary technique. Van Gogh had a vivid sense of colour and was keenly sensitive to atmosphere. You may see that in his ' Cypresses ' at the Tate Gallery, but his training was wholly inadequate. He may have felt and probably did feel intensely, but he only devoted himself to painting during the last eight years of his life, and finally went off his head and committed suicide. The only justification of the interest in his work is the tragedy of his failure. That is to say his reputation rests not on what he did, but on what critics have said he was trying to do. He is converted into a stalking-horse for incompetence. To do Van Gogh justice he possessed one essential quality of an artist, an almost passionate sensibility to nature, undaunted courage in trying to win from her the secret of her charm, and he is on a very different footing from those who turn their back on nature or hope to cut the knot by omitting all that it is difficult to draw. My second instance is taken from a recent book of travels, in which the writer's abundance of adjectives is balanced by a complete absence of drawing in the illustrations. These are so abstract as to be meaningless, without what on the Shakespearean stage were called the 'nuncupations', announcing this is a tree, this a woman, that a cow, and so on ! I recollect one illustration in particular, in which the faces of some figures in the foreground were indicated by white pear-shaped patches. After all, graphic art must make its appeal through visual images, and when the representation is reduced to a point of abstraction at which it ceases to resemble the actual phenomena of the visible world, one feels that Art has forgotten its purpose, that in fact this is not Art at all, but gibberish.

Unkind people regard these movements, which succeed each other with startling rapidity, Impressionism, Cubism, Significant form, Non-representationism, and the deliberate avoidance of form of any significance whatever, as simply so much advertisement. I believe myself that they sprang in the first instance from impatience with accepted methods, from the legitimate ambition to go farther, which is the necessary basis of all advance, but the pendulum of reaction has swung too far. In the revolt against authority and in their anxiety to assert themselves at all costs, the iconoclasts have forgotten what has actually been done. They have no real standard of attainment, nothing by which to measure the value of their own achievements ; and they are encouraged in this rake's progress by writers who talk of 'adventure' and make the test of Art the enjoyment with which the work has been produced, and who seem to think that if a work of art is unintelligible, it must conceal an idea of profound significance which it is the critics' business to disinter, and it must be confessed that when a picture which looks like an old-fashioned kaleidoscope is entitled ' the Creation of man ' there is indeed room for the ingenuity of the critic. Adventure is very well, but one does not start on a voyage of adventure with a half-trained crew and without a rudder. And enjoyment is very well; many an amateur has enjoyed his own labours, but he would be rash to assume that the fact of that enjoyment gives any particular value to his work.

One has noticed with regret, since the war, a disinclination for the discipline of apprenticeship, both in the Arts and in Literature. Jack is as good as his master, or rather better, and his idea of showing that this is so is to turn his back on accredited methods, ignore the past, and shout his crude ideas in some unintelligible jargon of his own. Surely this is hardly the way to repair the enormous intellectual ravages of the war. There are still the old ways which are the better, and there are no short cuts. The need for intellectual discipline is not less but greater, if we are to win back all that we lost in the war, if we are to recover the clear vision of beauty that inspired the work of the great masters of the past.

What we suffer from is, I think, confusion of ideas as to the meaning and purpose of Art; and one has to admit that this confusion is of very old standing. Plato's attitude to Art, for example, has always been a difficulty. Are we really to accept his contemptuous dismissal of Art as an insignificant affair, thrice removed from reality? Does the art of painting mean no more than a simulation of natural objects so exact that the picture might be mistaken for the object painted? That, strangely enough, appears to have been the view of the art held by the Greeks of Plato's time, and Plato the philosopher accepted it as true, and without further consideration relegated the Arts to a servile position, and dismissed them as injurious to the training of the young citizens of his ideal State. He even dismissed the Poet from his State, on the ground that he represented bad as well as good, and though he would have anointed him with myrrh and set a garland of wool on his head, he would none the less have sent him out into the wilderness to die. It seems curious to us, living in a far more complex civilization, that Plato should have overlooked the educational value of beautiful surroundings. Had he ascended the Acropolis and even glanced at the Parthenon, one would have thought that he would have felt the inadequacy of his own theory. Perhaps when the world was younger beauty was taken for granted, as part of the natural order of things. In any case the Arts were nothing but a side issue to Plato. Himself one of the most consummate artists in words that has ever lived, he seems to have been indifferent to other forms of Art. His aim was ethical, and he was convinced that an external object has only a partial, temporary, and incomplete existence, as compared with the Idea, the archetype conception of the object as it exists in the pure reason. He had therefore little difficulty in showing that the representation of these objects was at a lower level still, and that the practice of the graphic and plastic arts is immoral and dangerous. At first sight Plato seems of little use to us, in our search for the meaning of Art. One says it with diffidence in speaking of one of the master minds of the world, but I cannot help thinking that Plato and his followers took a careless view of Art, and did not even carry home their own philosophy in dealing with it. They dismissed it as an imitation of an imitation and declined to discuss it further. Plotinus the famous Neo-Platonist refused to have his portrait painted, because, he asked, 'Is it not enough to have to bear the image in which nature has wrapped me, without consenting to perpetuate the image of an image.' How different the history of aesthetic might have been if the Platonists had realized that Art is not a mere monkey-trick of imitation, but a search for ideal beauty, the sincere effort to disentangle the universal from the particular, the permanent from the accidental and the ephemeral. I believe myself that unless one accepts the view that beauty is absolute and not relative, or in terms of Platonism that there is an Idea of beauty, there is nothing in front of the Arts but bankruptcy. On the face of it Plato's theory of Art is most disappointing. Yet I believe that indirectly his theory of the Idea has a vitally important bearing on the whole position of Art.

Aristotle devoted his criticism almost entirely to the drama and only referred to the Arts by way of illustration, but it is clear from his scattered remarks that he had a definite theory of aesthetic, and this theory was a great advance on the doctrinaire position of Plato. In the first place, he separated the Arts from Ethics, and instead of treating them as negligible assigned them a definite place and function in the scheme of life. Their object, he held, was to produce pleasure in the spectator, partly emotional, partly intellectual. In a rudimentary form this pleasure would be given by the recognition of resemblance, by seeing at a glance that 'this is that',[1] that the picture or the model resembles some familiar object, and Aristotle had arrived at the advanced modern view that, even if the object represented is not pleasant, any representation that is well done will give pleasure, a view, by the way, not reconcilable with the Platonic position.

Further, it is the business of Art to find the universal in the particular, and present it in intelligible form to the spectator ; 'to image forth the immanent idea which cannot find adequate expression under the forms of material existence.' Butcher,[2] whom I have just quoted, puts it admirably in these words. By means of Art ' the pressure of everyday reality is removed, and the

---

[1] Ὅτι τοῦτο ἐκεῖνο, ὥστε μανθάνειν τι συμβαίνει, *Rhet.* i. 11, 1371, v. 4.

[2] S. H. Butcher's *Aristotle's Poetics*, p. 128.

aesthetic emotion is released as an independent activity'. For the first time Art was treated as a factor in life worthy of serious consideration. Yet Aristotle still approached it from the point of view of Ethics. He would have had no use for landscape art, and very little for sculpture unless it concerned itself with some heroic figure in action. His outlook on Art was limited absolutely by man—and as for architecture he was unable to place it anywhere except in the subordinate position of the merely useful and even servile. Moreover, he regarded the artist from the habitual standpoint of the philosopher, as a person without a right to his own independent existence. He could not conceive of a work of art as something complete in itself, apart from its effects on the spectator ; or of an artist creating for his own satisfaction, and claiming that his work should be judged on its own merits, apart from the effect it may have in directions external to itself. It is ominous that Aristotle makes, not the artist, but 'ὁ χαρίεις', the 'gentleman of taste and leisure', the final arbiter of the Arts ; and I always wonder that the gentlemen of taste, the Horace Walpoles and Comtes de Caylus of the eighteenth century, did not make more play with Aristotle's theory. The probable reason is that neither of these elegant noblemen had even heard of it. In the eighteenth century Du Fresnoy's *Art of Painting* still held the field, and according to Diderot the Comte de Caylus could not translate his own collections of inscriptions.

Throughout antiquity the work of the painter, the sculptor, and the architect seems to have been taken for granted. People did not discuss it or even think about it, because they could not conceive of it as other than it was. And it seems to have been much the same in medieval times. A few names of master builders have reached us, yet we know little or nothing about them. My impression is that the artists and the craftsmen were still regarded as men of little account. The change came with the Renaissance. Up to that date artists had been humble persons working among their fellows and with their fellows, on the same plane, and without any claim to exceptional merit. With the Renaissance they emerge as individuals. That movement started from above, from the scholars and intellectuals, and worked downward till it reached the artist and the craftsman. The noble patron appears upon the scene, to run his artist like a racehorse against his rival's entry, the Arts became conscious of themselves, and henceforward the initiative of the individual was to take the place of the predestined course of immemorial tradition, a process which has ended in the alarming results of latter-day Art. It was inevitable that this consciousness of Art must lead to disquisitions on its nature and purpose. In the first instance, these were almost entirely technical, and consisted of expositions of the practice of the ancients, as laid down by Vitruvius, and exemplified in the ruins of ancient Rome. About the middle of the seventeenth century Du Fresnoy, a not very successful painter, brought out a didactic Latin poem on the principles of painting, which was translated by Dryden, and again in the eighteenth century by the Rev. William Mason with notes by Sir Joshua Reynolds ; but neither Du Fresnoy nor his commentators made any attempt at a philosophy of Art, and Du Fresnoy's poem opens with a fatal adaptation from Horace, ' Ut Pictura Poesis erit, similisque Poesi sit Pictura ', an adaptation which provided the text of Lessing's *Laocoon*, by far the finest piece of constructive criticism written since the Poetics of Aristotle. If Aristotle rescued the Arts from Ethics, Lessing no less surely marked out their boundaries and limitations. Except that Hegel extended the range of Art to all nature, or as he might have put it to God as immanent in the universe, I cannot find that he made any very material advance, but I confess to finding Hegel's aesthetic exceedingly difficult to follow ; and Signor Croce's lucubrations on the graphic and plastic arts seem to me to leave them pretty much where they were. Always excepting Aristotle and Lessing, the philosophers are too profound to illuminate the practice of the humble working artist. Indeed they do not really concern themselves with the artist at all. To them he is still the necessary servile person. Their interest in the Arts is limited to the effect works of art may have on the spectator. It is purely subjective, and though in his candid moments the philosopher would have to admit the necessity of the artist, he takes his work for granted, and only accepts it as the necessary condition of certain states of feeling aroused in himself.

Yet the working artist ought to be able to give some account of himself, some indication of what goes on when he conceives his work of art and carries it out, of what he aims at and of the means by which he makes his appeal to the spectator. Regarding the creation of a work of art as a whole, I think we shall find that the mental process in the arts of painting, sculpture, and architecture is practically the same, but we have to differentiate between them in the early stages; that is to say in regard to the occasions which bring the activity of the artist into play. The landscape painter, for instance, starts with his intuition of external objects, the impact of some visible object on his sensibility through the sense of sight, such as the play of light and shade on a landscape, the movement of wind-swept trees against the sky, or the action of some figure. In other words, the first impulse comes from without; but in decorative art, in sculpture (excepting portraiture) and more especially in architecture, the process is rather from within outwards; that is to say, the mind starts with certain definite intentions, certain specific problems to solve; and though contact with externals must be established before any realization is possible, the externals are not the actual starting-point of the creative process. Architectural design in its initial stages is peculiar to itself. In the first place there are very definite limiting conditions, the laws of statics and dynamics, the properties of materials, the specific programme of the building to be designed, and whereas the painter and the sculptor from the first have to keep their work in touch with visible objects, the architect works by symbols; that is to say, the drawings through which his conceptions materialize are not representations of objects, but are symbols of them. But in the succeeding stages the Arts have this common ground, that a creative activity is set going in the mind of the artist, whether painter, sculptor, or architect; whereas in the non-artistic person no activity would be set going at all. To Peter Bell

> A primrose by the river's brim
> A yellow primrose was to him,
> And it was nothing more.

Whereas to the artist it would be everything. By his nature he is keenly alive to the visible world around him, it rouses in him the aesthetic emotion, and impels him to translate some part of it into a version of his own, whether his object is to satisfy himself or to convey his own intense feeling to others. What takes place is a process of transmutation. The *causa causans*, the jumping-off point if I may so call it, may be external nature, the subject before the artist, or the programme set him; but when the creative activity is thus aroused, what we vaguely call the mind works in and around the subject, till, in a flash as it were, the right realization reveals itself. All along, it would seem, some part of the mind has been working subconsciously, at some point the conscious and the subconscious meet like flint and steel, the fire kindles, the vision is realized, and the problem solved. This, put quite crudely, is more or less what happens when a work of graphic or plastic art is produced; and the essential conditions are the possession of exceptional powers of observation and imagination—the artist is the rare exception, not the rule; and in the second place, the possession of technical ability adequate at least to the free and unhampered expression of the idea. The idea and its expression cannot be separated. It is no use saying the idea was good but the expression bad. It is the excellence of both in complete fusion that makes the result a real work of art.

There is one further point in regard to the arts of painting and sculpture. In both cases the activity of the artist has reference to visible objects and he can only think in terms of visible objects, can only express the emotions aroused in him by means of the representation of these objects of sufficient similitude to be recognized as such. The mistake of certain recent methods in Art is that nobody can form any idea of what it is all about; an artist who appeals through the representation of visible objects should not require an interpreter to explain what he means. On the other hand, the artist is not an illusionist. The facts with which he deals are the starting-point of the idea that shapes itself in his mind into an actuality of its own ; an actuality which is indeed based on what has been transmitted to his mind through the eye, but which through the transmutation of the artist's mind acquires an independent existence of its own. Though based on an impression, and though the impression is an essential element, the finished work is more than an impression, it is or should be the complete statement of the result of a complex process, involving the impression or programme plus the mental transmutation, and throughout it all, the personal equation of the artist himself and his technical ability. It is this personal equation after all that gives its ultimate value to a work of art, and makes us prefer one man's work to another for its vision, its selections, its reticence on the one hand, its emphasis on the other, for the range of imagination that it reveals, for the glimpse that it gives us of the working of some exceptional spirit. The distinction of a work of art comes from within. It is not a matter of style planted on from without, but the result of the inner working of an individual soul. It is in this sense that there is no room in Art of any kind for the vulgar and the commonplace; whatever it may have been in the past, Art nowadays must be essentially individual.

If this account of the nature of graphic and plastic art is accepted, certain conclusions follow. In the first place, photography is eliminated from the province of Art, because it is a mechanical process, and because we cannot get from the photograph any idea of the impression that the objects represented have made on the mind of the photographer. Illusionist art, pictures that aim at making one believe that the representation is actually the object painted, or waxwork figures which tempt us to think that the eminent statesman or murderer stands before us in the flesh come under the same condemnation. Cubism and Futurism and the various other crazes of modern painting represent the opposite extreme, and appear to me to be almost as far removed from real Art. When painting or sculpture is the medium of expression the appeal is made through the eye, and, if it is to come home, it must do so in recognizable terms. We are familiar with those extraordinary productions in which the figures are like spillikins, the faces resemble deal boards, the rules of perspective are ignored, and the colours entirely arbitrary. These works are described as 'personal', 'grimly humorous', and so on. The painter seems to be obsessed with the notion that it is his business to give the abstract idea of his subject as he conceives it; but when he reduces it to a point of abstraction which ceases to be recognizable his justification is gone. He gives us no pleasure by form or colour, and the idea which he attempts to formulate would be far better expressed by the written word. His work is one more instance of the danger of disregarding the limitations of the Arts. What may be done by the written or spoken word cannot always be done by means of graphic or plastic art, and what can be done in sculpture cannot always be done in painting.

There is a third conclusion to be drawn from our premises. In every work of art one looks for personality, for some definite contribution by the artist beyond merely mechanical skill; and this is why Art is so difficult. 'No man by taking thought can add one cubit to his stature', and without the seeing eye, the quick sensibility, the free ranging imagination of the artist born, Art is, I believe, out of the question. A hundred years ago Benjamin Haydon deliberately set himself to produce great historical paintings, hoping, by the study of anatomy and of the Elgin marbles and the works of Raphael, to produce a work that should combine the merits of all. But in spite of some ability and an almost frenzied enthusiasm, his Dentatus and his Lady Macbeth are forgotten, and, like Van Gogh, poor Haydon's efforts ended in disaster and suicide. 'He who uninspired by the Muses comes to the doors of the Palace of Art convinced that by mere craftsmanship he will become a worthy artist, fails of his purpose both he and his art, commonplace art has been killed by the art of the inspired.'[1]

I have been reading lately Longinus on the 'Sublime', one might almost say 'the grand manner'. Longinus, a Greek writer of the first or third century, entertained a profound admiration for the mighty men of old, Homer in particular; and a contempt no less profound for the unseemly mannerisms of his own time, which he attributed to what he called 'a Corybantic craze for novelty', and he suggested that aspiring writers might do well to ask themselves three questions. In the first place, given the theme, how would Homer or Plato or Thucydides have handled it? In the second place, when the work was completed what would Homer have thought of it? And in the third place, how will it strike posterity? Some of our latest Corybantic performers might well ask themselves similar questions, for underlying their performances is the dangerous theory that beauty is relative, that there is no such thing as an abstract Idea of beauty, if I may so use Plato's phrase. It is only on the hypothesis that one thing is as beautiful as another, if only a man is found to think so, that it is possible to explain

[1] Ὅς δ' ἂν ἄνευ μανίας Μουσῶν ἐπὶ ποιητικὰς θύρας ἀφίκηται, πεισθεὶς ὡς ἄρα ἐκ τέχνης ἱκανὸς ποιητὴς ἐσόμενος, ἀτελὴς αὐτός τε καὶ ἡ ποίησις ὑπὸ τῆς τῶν μαινομένων ἡ τοῦ σωφρονοῦντος ἠφανίσθη, Plato, *Phaedrus* 245 A.

the deliberate departures from all standards recognized in the past. That the result in the Arts must be chaotic and end in their dissolution, our Corybantics do not stop to think.

We come back to the parting of the ways. Those who hold that beauty is merely relative very soon come to the conclusion that there is no such thing at all; they delete it from their programme, and the results are seen in the exhibitions of what purports to be modern art in some of our minor galleries. For myself, I take my stand on the Platonic Idea, and in the words of the Dean of St. Paul's, I venture to think that ' Greek sculpture is absolutely beautiful, while Cubist art is intrinsically and objectively hideous.' This does not mean any rigid formula of art or canon of beauty; it means that beauty differs from ugliness as light differs from darkness, and that for all of us artists there exists an unapproachable and unappeasable ideal of beauty that we can never reach completely. The history of Art seems to me to be the history of man's imperfect effort to attain to that ideal of beauty. Sometimes, and very rarely, the genius of man has realized the perfect standard, in the Doric of the Parthenon for instance, the Aphrodite of Cyrene, or the victory of Samothrace, or in the third Brandenburg Concerto, Mozart's clarionet Quintet or the last movement of the Choral Symphony. These are the rare revelations that the genius of man is almost illimitable, and that still keep alive the unconquerable hope ; but no one knows better than the artist himself how far his work falls short of his inner vision.

> Donec longa dies, perfecto temporis orbe,
> concretam exemit labem, purumque reliquit
> aethereum sensum, atque aurai simplicis ignem.[1]

Yet the work of interpretation, of conveying to others this lofty vision, is surely a noble one. It is not for the artist to cut capers to the pit, or play for the applause of the gallery. Rather it is his high privilege to give the finest expression that he can to the thought and emotion within him, and in doing so let him think of an audience beyond the reach of advertisement and intrigue. For all he knows the spirits of the mighty dead may be watching him, and far into the distant future stretch the ranks of the generations to come. If his work has any element of greatness in it, somewhere among those ranks will be found a kindred spirit, and his appeal will not have been made in vain.

[1] *Aen.* vi. 745–7.

June 1924.

www.ingramcontent.com/pod-product-compliance
Lightning Source LLC
Chambersburg PA
CBHW020227170426
43201CB00007B/341